AF587849

CELL BIOLOGY RESEARCH PROGRESS

CACO-2 CELLS AND THEIR USES

Cell Biology Research Progress

Additional books in this series can be found on Nova's website
under the Series tab.

Additional E-books in this series can be found on Nova's website
under the E-books tab.

CELL BIOLOGY RESEARCH PROGRESS

CACO-2 CELLS AND THEIR USES

MEGAN A. SCHULZ
EDITOR

Nova Science Publishers, Inc.
New York

For permission to use material from this book please contact us:
Telephone 631-231-7269; Fax 631-231-8175
Web Site: http://www.novapublishers.com

Library of Congress Cataloging-in-Publication Data

Caco-2 cells and their uses / editor, Megan A. Schulz.
p. ; cm.
Includes bibliographical references and index.
ISBN 978-1-61209-917-0 (hardcover : alk. paper)
1. Epithelial cells. 2. Monomolecular films. 3. Intestinal absorption.
I. Schulz, Megan A.
[DNLM: 1. Caco-2 Cells.
2. Intestinal Absorption--physiology. QU 300]
QP88.4.C33 2011
611'.0187--dc22

2011004575

Published by Nova Science Publishers, Inc. † New York

CONTENTS

PREFACE

CACO-2 cell cultures are widely used in vitro models in several scientific fields such as nutrition, physiopathology, pharmacology, and toxicology. When cultured onto filters, these cells form a highly polarized epithelium that allows discrimination between apical and basolateral compartments. This new book presents topical research in the study of CACO-2 cells and their uses. Topics discussed include the application of CACO-2 cell models as a supportive in vitro tool during drug discovery and formulation; co-culture systems with CACO-2 cells to evaluate the bioactivity of plant extracts and the use of CACO-2 cells in intestinal transport, cell-matrix interactions and wound healing in intestinal villus cells. (Imprint: Nova Press)

Chapter 1 - The application of CACO-2 cell models as a supportive *in vitro* tool during drug discovery and formulation development aims to address the per oral absorption of drug candidates with different degrees of predictive efficiency, depending on the inherent CACO-2 cell characteristics and on the intra-laboratory quality control protocols. Although CACO-2 cell permeability studies probe a certain subset of experimental questions, careful interpretation of the obtained data should be exercised on more complex issues, which include the interplay between multiple biological functions. In the first part of the presented paper the academic applicability of CACO-2 cells for investigating drug-food interactions arising as a consequence of modified transporter activities will be demonstrated. To illustrate these claims, the suitability of the CACO-2 cell monolayers as a reliable instrument in predicting the absorption will be manifested first. Then the impact of garlic supplement (aged garlic extract) on the permeability of selected cardiovascular, antidiabetic and antiviral drugs will be presented and compared to the permeability determined through the isolated segments of a rat's small intestine. Based on the provided review regarding the (dis)similarities in the transporter and enzyme expression between this cancerous cell line and intestinal tissue, the observed qualitative pharmacokinetic transporter-drug interactions will be considered in the context of their *in vivo* relevance. The second part of the paper will focus on the industrial interest in CACO-2 models as means to evaluate novel drug delivery systems. Development of adequate pharmaceutical formulation is the valuable key step for improving poor water solubility and/or poor membrane permeability that are responsible for the low fraction of the drug dose absorbed. Modern approaches to address this problem include administration of the drug with lipid-based systems. Model drug furosemide, classified as a low permeable and a low soluble drug, will be used to identify beneficial properties of these novel drug delivery systems that could help overcoming the absorption hindrances and offer new possibilities for oral administration of other compounds with unfavorable physical and/or chemicals properties.

Chapter 2 - In nutrition research, *in vitro* cellular models have been considered as an expedient approach to the study of biological mechanisms underlying the effect of molecules of nutritional interest on different physio-pathological contexts and as the method of choice to obtain observations that can be extrapolate to *in vivo* studies. However, some critical points still need to be elucidated to land to definitive conclusions on the applicability of specific cell lines for the understanding of mechanisms associated with physiological processes. Although significant progress has been made throughout the time, it is generally recommended to take experimental results on cultured cells wisely and with caution. This chapter outlines some aspects related with the useful of cell lines in intestinal absorption studies, including the identification of the experiment's objective, molecular mechanisms and biological aspects involved at cellular level and the interpretation of results that can be elucidated by *in vitro* assays, including approaches recently developed. The authors will mainly focus on specific cell lines used as a model for intestinal absorption, discussing the "pros and cons" and the critical aspects to be considered while planning a study. The value and possible limitations of the utilization of some specific cell lines will be presented and discussed, commenting on some aspects of the clinical predictive value of these *in vitro* models.

Chapter 3 - Natural extracts are potential candidates for the development of new functional foods, but these extracts are usually complex mixtures of molecules of uncertain bioavailability and are often partially metabolized before reaching the target cells *in vivo*. Therefore, *in vitro* studies of the bioactivity of these extracts in cell culture may not translate to real-life applications *in vivo*. To overcome this limitation, co-culture systems that simulate the intestinal barrier using CACO-2 cells have been developed. Human CACO-2 enterocytes seeded onto culture inserts can be co-cultured with the cell type in which the plant extract effects are to be analyzed. Using this system, plant extracts applied to the apical side of CACO-2 cells have to be absorbed and metabolized by the cell barrier to reach the cells of interest. The system has been tested with different cell types such as pancreatic rat beta cells (INS-1) and human hepatocytes (HEPG2). CACO-2 cells in co-culture maintain their abilities to filter and metabolize the tested plant extracts. In the same way, key functions of the cell type of interest (i.e. triglyceride secretion in hepatocytes and glucose-stimulated insulin secretion in INS-1) are not affected by the co-culture. Therefore co-culture of CACO-2 cells with other cell types seems to be a suitable method to evaluate the bioactivity of natural plant extracts. However, there are some aspects that must be carefully considered before generalizing the appropriateness of this method. First, the co-culture may induce some adaptations in one or both cell types. For example, in INS-1 and HEPG2 co-culture systems, gene expression of the glucose transporter SGLT-1 is modified. Thus it should be considered whether adaptations may affect the function of interest and tested for. Furthermore, long-term co-culture can influence the viability of cells. Therefore, the time frame in which the cells are normally functional must be established for each cell type. As an example, HEPG-2 cells can be co-cultured for at least 24 hours without effects on cell viability while INS-1 co-culture studies should be restricted to shorter times (i.e. 4 hours).

Chapter 4 - CACO-2 cell cultures are widely used *in vitro* models in several scientific fields such as Nutrition, Physiopathology, Pharmacology, and Toxicology. When cultured onto filters, these cells form a highly polarized epithelium that allows discrimination between apical and basolateral compartments. For this reason, CACO-2 cells constitute a suitable experimental strategy for the study of substrate transport across either the transcellular or paracellular pathways of the intestinal epithelium. Under pre-established experimental conditions, these cultures can be used to estimate intestinal absorption (apical-to-basolateral fluxes) and secretion (basolateral-to-apical fluxes). Moreover, they are also highly used for the functional and molecular characterization of the transport systems specific to the apical and basolateral membranes of the enterocyte. Furthermore, dietary regulation of intestinal transport can also be examined by introducing changes in the composition of the apical compartment. Finally, the paracellular pathway, through tight junctions, can be studied using specific substrates of this route. This chapter discusses the value of experimental approaches using CACO-2 cells to address intestinal absorption.

Chapter 5 - The human colon carcinoma cell line CACO-2 has been widely characterized and used to investigate intestinal cell function and barrier regulation. These cells spontaneously undertake a complete intestinal-like program of epithelial differentiation, but their cancerous nature, as well as their colonic origin, has lead to reservations as to the physiological relevance of these cells. However, numerous studies have demonstrated that CACO-2 cells share many characteristics with the differentiated villus cell population of the normal intestine. This review focuses on the use of CACO-2 cells for the study of cell-matrix interactions and wound healing processes in intestinal cells. Rapid and efficient healing of epithelial damage is critical to the functional integrity of the small intestine. Several studies have reported the ability of CACO-2 cells to synthesize and secrete extracellular components that can affect the capacity and rate of epithelial restitution. Moreover, epithelial repair is a complex process relying on the coordinated action between epithelial cells and the underlying contractile cells. The authors have described a fully polarized CACO-2 cell monolayer grown on top of human intestinal myofibroblasts as a co-culture system allowing the deposition of a physiologically relevant basement membrane – like structure resulting from the mutual inductive effects of each cell type on the other. Monocultures of either cell type do not generate a basement membrane, but are only capable of forming a primitive insoluble matrix. Monitoring of wound healing showed drastic morphological changes in epithelial polarized columnar cells at the edge of the wound and repair was significantly more efficient in CACO-2/myofibroblast co-cultures. Taken together, these results indicate that this co-culture system is a valid tool to investigate the mechanisms underlying human intestinal epithelial restitution as well as the key players involved in their regulation.

Chapter 6 – The authors' aim was to investigate the effect of *n*-3 PUFAs (docosahexaenoic acid (DHA) and eicosapentaenoic acid (EPA)), *n*-6 PUFAs (linoleic acid (LA), γ-linolenic acid (LNA) and arachidonic acid (AA)), conjugated linoleic acid (CLA) and the bile salts deoxycholic acid (DCA) and taurodeoxycholic acid (TDCA) upon uptake of ^{14}C-butyrate (^{14}C-BT) by intestinal epithelial cells. CACO-2 cells were exposed to LA, LNA, AA, DHA, EPA, CLA and DCA (100 μM) for 1 to 7 days. The only significant effects were slight (7-18%) increases in ^{14}C-BT uptake found with EPA, CLA, AA and DCA. IEC-6 cells were exposed to LA, LNA, AA, DHA, EPA, CLA, DCA and TCDA (100 μM) for 2 days. LNA, EPA, CLA and DHA caused small (9-16%) decreases in ^{14}C-BT uptake. For CLA, this effect was associated with a decrease in cellular viability. A discrete interference of *n*-3 and *n*-6 PUFAs, CLA and unconjugated bile acids with the apical uptake of BT was observed.

Chapter 7 - Food-borne and intestinal (infant and adult) botulism are the most common forms of human botulism, and both result from the passage of botulinum neurotoxin (BoNT) from the digestive tract into the circulation. Interestingly, BoNT retains its toxicity after passage through the digestive tract despite being a large protein toxin (approximately 150 kDa). BoNTs are produced as large protein complexes (BoNT complex) associated with non-toxic components, including non-toxic non-hemagglutinin (NTNH) and hemagglutinin (HA). These proteins protect BoNT from the acidic and proteolytic environment of the digestive tract, and, moreover, recently showed that HA disrupts the intercellular epithelial barrier by directly binding to E-cadherin. As an *in vitro* human intestinal epithelial barrier model, the CACO-2 monolayer system is among best experimental systems to advance these studies. This chapter provides an overview of the recently elucidated molecular interactions between the BoNT complex and the intestinal epithelial barrier.

Chapter 8 - The strategy to estimate the genotoxicity of xenobiotics includes *in vitro* assays detecting mutations or primary DNA lesions. Among the various cell lines that can be used, the intestinal ones are relevant for the study of food contaminants as the first barrier to cross to reach the systemic circulation. The human CACO-2 cells is a common model for toxicokinetics and numerous publications indicate specific activities (influx and efflux transporters, metabolic enzymes) which can play a key role in the toxicity of food contaminants. *In vitro* genotoxicity studies have been recently performed on this cell line and most of them are reported in this review. Some improvements to overcome the limits which can be encountered with the CACO-2 cell line can also be proposed depending on the aim of the study. As examples, co-cultures with other cell types as well as regulating metabolic activities can be considered.

Chapter 9 - CACO-2 cell model has evolved as a useful tool for studying human oral drug absorption. The model is suitable for a highthrougput system, in order to rank compounds in the screening phase, but it could be also used for mechanistic studies by adapting the experimental conditions, for instance using different initial concentrations and performing bidirectional experiments. With the adequate experimental design and data treatment it allows to explore the permeation mechanism (i.e passive versus carrier mediated) and to evaluate the transport parameters. In this chapter the use of CACO-2 cells for prediction human oral fraction absorbed will be discussed as well as the advantages and disadvantages versus other in vitro models (as PAMPA membranes) or in situ models as the rat perfusion method. In second place, the different mathematical approaches to obtain transport parameters from the CACO-2 model will be described and discussed in terms of experimental burden, accuracy and relevance for the aim of the study. Finally several examples of mechanistic studies of fluoroquinolones will be used to illustrate the detection of secretion mechanism and the correlation with the results obtained in rat. Examples will be provided from our experimental experience on the study of fluoroquinolones but results from other papers will be also reviewed and discussed. This short review does not intend to be a comprehensive compilation of all the CACO-2 related information and published experiments but to point out some useful examples in order to help the users to optimize the use of this model.

Chapter 10 - Caco-2 cells are widely and commonly used to study oral absorption of drugs. Colloidal drug delivery systems, such as nanoparticles, nanocapsules, liposomes, micelles, have been proposed to enhance the absorption of poorly absorbed and poorly soluble drugs (BCS class IV). Caco-2 cells can be used to study, compare and screen the performances of these drug delivery systems. In this chapter, experimental studies with nanocarriers and Caco-2 cells will be described and analyzed. Two different types of studies will be considered: absorption studies (apparent permeability determination though cell monolayers under various experimental conditions) and mechanistic studies (determination of the main routes of uptake by the cells and the ratio between passive and active pathways). For each type, examples will be given and analyzed in order to shed light on the envisioned findings obtained with Caco-2 cell tool but also on the pitfalls of these studies. In a last part of the chapter, Caco-2 cell monolayers, will be compared with other cellular models and biopharmaceutical tools such as PAMPA's or Single pass perfusion model.

Chapter 11 - Diets rich in fruits and vegetables are associated with a lower risk for colorectal cancer (CRC). Fruits and vegetables contain procyanidins, oligomers of flavan-3-ols that are among the most abundant polyphenols in human diets. Large procyanidins, i.e. composed of more than 3 monomer units (LPCA), are not transported inside cells. However, LPCA could act at the gastrointestinal level preventing chronic inflammation and the associated increased risk to develop CRC. CACO-2 cells is the model of intestinal epithelium that were used to study the mechanisms through which LPCA could act protecting the integrity of the intestinal epithelial barrier, preventing oxidation and the triggering of inflammatory signals. It was observed that LPCA bind to the plasmatic membrane and protect CACO-2 cells from oxidants, from the cytotoxic effects of deoxycholic acid (DCA), a secondary bile acid, and from inflammatory molecules, i.e. tumor necrosis alpha (TNFα). Mechanistically, it is plausible that LPCA act through: a) their free radical scavenging capacity; b) their direct interaction with membranes lipids, causing changes in membrane biophysics that would affect the activity of membrane-associated processes; and c) their interaction with membrane receptors that trigger intracellular responses. In summary, CACO-2 cells have proved to be an excellent model to study the beneficial health effects of the non-absorbable LPCA. These findings further support the use of CACO-2 cells to identify the mechanisms of the intestinal actions of food components and/or pharmacological agents.

In: CACO-2 Cells and their Uses
Editor: Megan A. Schulz, pp. 1-30
ISBN 978-61209-917-0

Chapter 1

THE INDUSTRIAL AND ACADEMIC RATIONALE OF CACO-2 CELL MODELS TODAY

***Katja Berginc*[*1], *Alenka Zvonar*[2], *Mirjana Gašperlin*[2] *and Albin Kristl*[1]**

[1]The Department for Biopharmacy and Pharmacokinetics, Faculty of Pharmacy, University of Ljubljana, Aškerčeva 7, 1000 Ljubljana, Slovenia

[2]The Department for Pharmaceutical Technology, Faculty of Pharmacy, University of Ljubljana, Aškerčeva 7, 1000 Ljubljana, Slovenia

ABSTRACT

The application of CACO-2 cell models as a supportive *in vitro* tool during drug discovery and formulation development aims to address the per oral absorption of drug candidates with different degrees of predictive efficiency, depending on the inherent CACO-2 cell characteristics and on the intra-laboratory quality control protocols. Although CACO-2 cell permeability studies probe a certain subset of experimental questions, careful interpretation of the obtained data should be exercised on more complex issues, which include the interplay between multiple biological functions.

In the first part of the presented paper the academic applicability of CACO-2 cells for investigating drug-food interactions arising as a consequence of modified transporter activities will be demonstrated. To illustrate these claims, the suitability of the CACO-2 cell monolayers as a reliable instrument in predicting the absorption will be manifested first. Then the impact of garlic supplement (aged garlic extract) on the permeability of selected cardiovascular, antidiabetic and antiviral drugs will be presented and compared to the permeability determined through the isolated segments of a rat's small intestine. Based on the provided review regarding the (dis)similarities in the transporter and enzyme expression between this cancerous cell line and intestinal tissue, the observed qualitative pharmacokinetic transporter-drug interactions will be considered in the context of their *in vivo* relevance.

The second part of the paper will focus on the industrial interest in CACO-2 models as means to evaluate novel drug delivery systems. Development of adequate

*Corresponding author: (Tel.: + 386 1 476 95 00; Fax: + 386 1 425 80 31; e-mail address: katja.berginc@ffa.uni-lj.si)

pharmaceutical formulation is the valuable key step for improving poor water solubility and/or poor membrane permeability that are responsible for the low fraction of the drug dose absorbed. Modern approaches to address this problem include administration of the drug with lipid-based systems. Model drug furosemide, classified as a low permeable and a low soluble drug, will be used to identify beneficial properties of these novel drug delivery systems that could help overcoming the absorption hindrances and offer new possibilities for oral administration of other compounds with unfavorable physical and/or chemicals properties.

Keywords: CACO-2, transporters, enzymes, interactions, garlic, SMEDDS, permeability, absorption.

ABBREVIATIONS

Pgp: P-glycoprotein
MRPs: Multidrug Resistance Associated Proteins
BCRP: Breast Cancer Resistance Protein
PepT1: Peptide transporter 1
OATP-B: Organic Anion Transporting Polypeptide B
MCT-1: Monocarboxylate transporter 1
OCTs: Organic Cation Transporters
OCTNs: Carnitine/Organic Cation Transporter
OATs: Organic Anion Transporters
GLUTs: Glucose transporter
HPT-1: endogenous oligopeptide transporter
ABC: ATP Binding Cassette transporters
SLC: Solute Carrier Transporters
BCS: Biopharmaceutical Classification System
DIDS: 4, 4′-diisothiocyanatostilbene-2, 2′-disulfonic acid
CHC: α-ciannohydroxycinnamic acid
PSC8333: Pgp inhibitor
MK571: MRP inhibitor
TRP-V: Vanilloid Transient Receptor Potential calcium channels
CFTR: Cystic Fibrosis Transmembrane conductance Regulator

1. INTRODUCTION

CACO-2 cells have been available as *in vitro* model to address a different subset of questions with regard to the drug absorption since 1977. Human in origin, CACO-2 cells were isolated from colo-rectal adenocarcinoma. Although they retain some undesirable colon characteristic (i.e. low paracellular permeability, the absence of mucus), they also exhibit pivotal morphological and functional attributes of the small intestinal epithelium (i.e. cell polarization, expression of some brush-border enzymes, formulation of microvilli), which develop during cell growth, differentiation and sometimes with the aid of manipulation protocols [1]. From the pool of different cell cultures, which have been used to study intestinal drug absorption (i.e. Madin-Darby canine kidney cell line MDCK, porcine kidney cell line LLC-PK1, rat fetal intestinal epithelial cells 2/4/A1, human colon cells co-cultured with mucous secreting cells HT29, mucin producing human colon adenocarcinoma cell line T84), the use of CACO-2 cells and their transfected (i.e. TC7) variants is still the most extensive [2].

The employment of CACO-2 cell line in the drug discovery process enables one to avoid numerous and complex whole animal studies for the identification of the most promising lead compound with desirable pharmacokinetic properties [3, 4]. However, newly synthesized compounds usually exert low and variable bioavailability associated with poor water solubility, poor permeability, extensive first pass metabolism and/or instability in the gastrointestinal (GI) lumen. In the attempt to circumvent these hindrances drug candidates are thus formulated as solid dispersions, nanoparticles and microparticles, or as formulations containing lipids, surfactants and permeation enhancers [5]. To justify the utilization of such formulation strategies, beneficiary properties of novel drug delivery systems, which help in overcoming undesirable physio-chemical or biological (metabolic and elimination) barriers, have to be demonstrated. To gain full insight into the mechanism that ensures sufficient and constant drug absorption from novel drug delivery systems, different *in vitro* intestinal cell models such as CACO-2 monolayers can be utilized [6]. Besides industrial interest in CACO-2 models, these cells have amply proven their worth also as *in vitro* models of many physiological functions (i.e. intracellular regulation, trafficking processes, and toxicity related issues) [7] and could also be used as a qualitative prognostic tool to evaluate the potential for pharmacokinetic drug-food interactions arising during drug absorption due to the competition between the applied drug and phytochemicals for mutual transporters, enzymes or both [8].

To address the above stated claims, the usefulness of CACO-2 cell models from industrial and research perspectives will be illustrated. This paper also provides a review, summarizing CACO-2 cell limitations, advantages and transporter/enzyme expression patterns. According to that and our *in vitro* results investigating drug-garlic interactions, the power of the CACO-2 *in vitro* model in assessing the probability of drug-food interactions will be demonstrated with regard to the *in vivo* relevance. Afterwards the beneficiary properties of self-microemulsifying drug delivery systems (SMEDDS) in furosemide oral drug delivery will be shown by establishing the underlying mechanisms aiding furosemide absorption with the help of the CACO-2 system.

2. CACO-2 Cell Advantages, Drawbacks and Transporter-Enzyme Expression Patterns

After reaching confluence, CACO-2 cells (Figure 1) differentiate into a polarized monolayer, losing typical characteristics of colonic tissue. Confluence ensures the beginning of the expression of transporters and enzymes and the development of typical morphology, similar to the one of enterocytes [9]. Based on the overflow data from different laboratories reporting transporter and enzyme expression patterns [10-13], marker substance permeabilities [1, 14], culture protocols [6] and transepithelial electrical resistances - TEER [15], one is prevented from direct data comparison due to the extremely high inter-laboratory variability. Establishing CACO-2 quality control tests, constant culture and experimental protocols however, ensures that CACO-2 retain their reliability and suitability as high-throughput screening *in vitro* model [9, 14, 16].

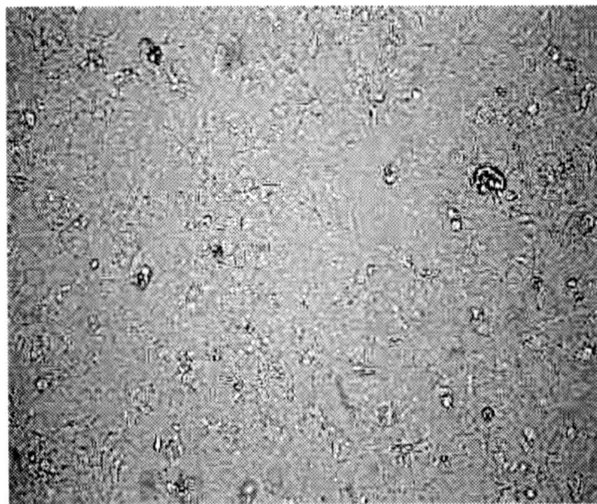

Figure 1. 3-day old CACO-2 cells (passage number 56) in culture reaching approximately 80% of confluence, needed before cell trypsinization and seeding onto filters.

2.1. Applicability of CACO-2 Models to Predict the Fraction of Drug Absorbed in Humans

According to the Guidance for Industry – Waiver of *in vivo* bioavailability and bioequivalence studies for immediate-release solid oral dosage forms based on biopharmaceutical classification system [17], CACO-2 monolayers are used for the prediction of the fraction of drugs absorbed (F_{abs}) from a human intestinal tract, if a correlation between the permeability of the suggested low (i.e. atenolol, furosemide, terbutaline) and high (metoprolol, naproxene, antypirin) permeable standard drugs, determined through CACO-2 cell monolayers, and human *in vivo* absorption data for the same standards would be established beforehand. Unfortunately, the static nature and the dissimilarities between the CACO-2 model and actual *in vivo* human physiology preclude the use of CACO-2 cells to predict F_{abs}, unless the investigated drug is a highly permeable compound, which transverses the membrane mainly passively and transcellularly [15]. When the studied substances are preferentially hydrophilic, their absorption is mediated by the uptake transporters or it involves permeation through the tight junctions (i.e. paracellular permeability). In these situations the predictive power of the CACO-2 model is low, because the expressions of the uptake SLC (Solute Carrier proteins) transporters, the number of tight junctions and their size are significantly lower than *in vivo* (the average diameter of tight junction opening in CACO-2 cells is ca 4 Å and ca 13 Å in a human small intestine, for SLC expression see Table 1) [3, 15, 18]. Furthermore, the additional *in vitro* intestinal permeability underestimation with CACO-2 monolayers for hydrophilic drugs comes from the fact that the absorptive surface area in CACO-2 cells is at least 30-times smaller than *in vivo*, because CACO-2 cells lack intestinal folds and villi. These structures entrap drugs and prolong their retention above the absorptive surface in the GI tract, which leads to either increased paracellular absorption through leakier tight junctions in the crypts or to a higher probability of interactions between the uptake transporters in the enterocytes and drug and consequently to a more profound drug uptake [4, 15].

The absorption of many compounds is often limited by the pre-systemic gut metabolism or enteric drug cycling mediated by ATP Binding Cassette (ABC) transporters and enzymes (i.e. CYPs), whose expressions and activities in CACO-2 cells show dissimilarities compared to human small intestine [15]. In order to correlate measured CACO-2 permeabilities with human data a thorough characterization of the extent of transporter/enzyme expression in the employed cell line has to be performed [3, 4, 15]. Additionally, the experimental conditions (i.e. composition of the transport solutions, pH, the maintenance of sink conditions, the prevention of non-specific drug binding to the plastics and filters - cacophilicity, sufficient recovery, stirring speed, temperature and atmosphere) have to be closely monitored [19, 20]. Although the above mentioned CACO-2 cell drawbacks prevent direct correlation of the measured permeability with F_{abs} values, the identification of compounds with potential absorption problems could successfully be projected, if low absorptive permeability of the studied candidate through CACO-2 cell monolayers could be obtained [15]. Namely, low apparent permeability (P_{app}) determined through CACO-2 cell monolayers for the investigated compound would directly indicate that the *in vivo* absorption of this drug would also be low due to either low passive (transcellular and/or paracellular) permeability or it would suggest the involvement of the active uptake by SLC transporters, which could aid attaining high *in vivo* plasma concentrations.

Following the Food and Drug Agency's Guideline for industry [17], the validation of CACO-2 cells by determining absorptive permeabilities of high (naproxen, verapamil, metoprolol) and low (furosemide, hydrochlorothiazide, atenolol, ranitidine) permeable reference drugs, tested at 0.1 mM donor concentrations, was performed in our laboratory and the permeabilities were correlated to F_{abs} values as shown in Figure 2. This allowed us to determine limit permeability values, below/above which a compound could be classified as low (P_{app} values below $1x10^{-5}$ cm/s) or high (P_{app} values above $5x10^{-5}$ cm/s) permeable [4, 20]. As indicated in Figure 2, two data clusters were obtained, one for high permeable and the other for low permeable compounds.

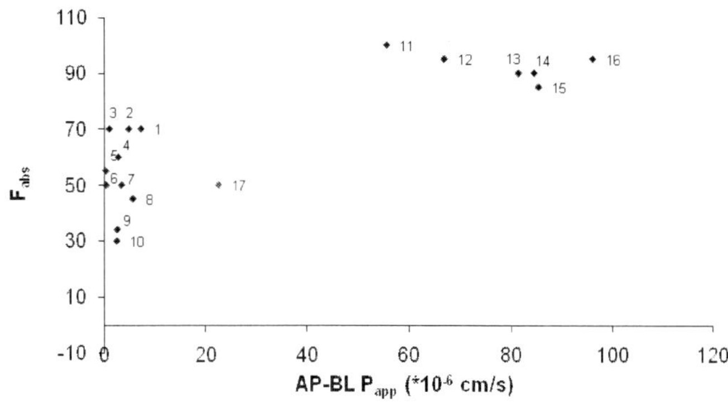

Figure 2. Correlation between permeabilities of drugs, determined through Snapwell® grown CACO-2 cell monolayers, and Fractions of drugs absorbed in vivo, according to the FDA guidance. 1 – digoxin; 2 – ciprofloxacin; 3 – hydrochlorothiazide; 4 – cimetidin; 5 – valsartan; 6 – furosemide; 7 – ranitidine; 8 – atenolol; 9 – pravastatin; 10 – atorvastatin; 11 – naproxen; 12 – verapamil; 13 – timolol; 14 – quinidine; 15 – lansoprazole; 16 – metoprolol; 17 - etopozide.

At least 85% of the administered dose has to be absorbed *in vivo* for a drug to be classified as highly permeable according to the Biopharmaceutical Classification Systems. If the drug does not meet this criterion, it is classified as low permeable [17]. F_{abs} values for naproxen, verapamil, timolol, quinidin, lansoprazol and metoprolol are all equal to or exceed 85% of the administered dose and their permeabilities were also higher than $5x10^{-5}$ cm/s. The group of low permeable compounds, whose permeabilities did not exceed $1x10^{-5}$ cm/s values, included atorvastatin, pravastatin, valsartan, furosemide, atenolol, ranitidine, cimetidine, hydrochlorothiazide, ciprofloxacin and digoxin. The only outlier was etoposide, which displayed "middle" permeability and has, according to the literature, highly variable bioavailability ranging from 25 to 75% [22, 23].

Although some of the compounds were classified as highly permeable they also displayed pronounced interactions with ABC (P-glycoprotein (Pgp) – lansoprazole, quinidine, verapamil, Multidrug Resistance Associated Protein (MRP) - verapamil) and SLC (Organic Anion Transporting Polypeptide (OATP) – lansoprazole, Monocarboxylic acid transporter 1 (MCT1) – naproxen) transporters *in vitro*. The participation of efflux transporters was confirmed in experiments with identical pH conditions on both sides of CACO-2 monolayers (pH 7.4), while the involvement of the uptake transporters was assessed with apical pH set to 6.5 and basolateral to 7.4. The relevance of transporters to *in vivo* absorption is usually negligible. Namely, for BCS I classified drugs (high solubility, high permeability: i.e. timolol, quinidine, verapamil) and some BCS II acidic drugs (low solubility, high permeability: i.e. naproxen, lansoprazole) per-oral therapeutic doses usually completely dissolve in 250 ml of concomitantly ingested water in the stomach (BCS I compounds) or in the duodenum (acidic BCS II drugs). High solubility in the stomach or duodenum and high permeability ensure sufficiently high intraluminal drug concentrations capable of rapid distribution into enterocytic membranes, which leads to the saturation of the uptake, secretory transporters or enzymes [24]. For other non-acidic BCS II drugs, the highest dose does not completely dissolve in the 250 ml of fluid. One could argue that these drugs formulated as immediate release dosage forms could completely dissolve in the acidic environment in the stomach, but it is known that after emptying the stomach content into the duodenum, the dissolved basic BCS II drugs encounter media with higher pH and start precipitating. The rate of drug precipitation could exceed the rate of absorption, leading to lower intraluminal concentrations than expected. These assumptions are correct in the majority of cases, where non-acidic BCS II drugs are involved, which has been substantiated by numerous non-bioequivalence studies for these types of drugs [25]. Following the above stated premise of low intraluminal concentration of non-acidic BCS II drugs, the impact of ABC transporters on the absorption of these representatives will still be significant, while the influence of SLC transporters will be minimal. Namely, although the attained intraluminal concentrations are not as high as in the case of BCS I/acidic BCS II compounds, the equilibrium between undissolved and dissolved drug and the drug's high lipophilicity will enable rapid drug distribution into mucosal membrane, which causes saturation of the uptake SLC transporters. However, the ABC transporters will still curtail absorption, since due to the low solubility in the cytoplasm of enterocytes (pH 7.4), attaining sufficiently high intracellular drug concentrations to saturate efflux transporters is limited. For other BCS III (high solubility, low permeability: i.e. pravastatin, ranitidine, cimetidine, atenolol, hydrochlorothiazide, valsartan) or BCS IV (low solubility, low permeability: i.e. furosemide, ciprofloxacin) classified drugs, uptake and efflux transporters and enzymes usually display a pronounced impact on their intestinal pharmacokinetics and can significantly modify the drug's plasma profile [24, 26].

Establishing correlation between *in vitro* determined permeabilities for the in-house CACO-2 cell line and human data (Figure 2) thus enables us to classify other investigated compounds as low/high permeable based only on the measurements of absorptive permeabilities through CACO-2 monolayers. Also, the involvement of efflux transporters in drug's absorption can be readily assessed with bi-directional permeability measurements through CACO-2 monolayers under isocratic pH conditions (i.e. apical and basolateral pH 7.4). The potential participation of the uptake transporters in drug absorption is usually re-evaluated with rat small intestine due to low uptake transporter expression in CACO-2. The experiments with rat intestine are performed in the same manner as with CACO-2 cells; the intestinal tissue (duodenum or jejunum - segments with the highest uptake transporter expression) is mounted in side-by-side diffusion chambers and exposed to the appropriate experimental conditions that have been shown to ensure the highest transport activities of the uptake transporters (i.e. acidic apical/mucosal pH, buffer ion composition...) [27]. The importance of passive paracellular permeability through tight junctions in the absorption of small hydrophilic and preferentially cationic molecules could be assessed by comparing the permeabilities determined through CACO-2 monolayers and PAMPA (Parallel Artificial Membrane Permeability Assay). Namely, permeation of drugs through PAMPA systems proceeds exclusively transcellulary, whereas through CACO-2 monolayers, transcellular and to some extent paracellular diffusion could be measured [28]. Alternately, paracellular diffusion could also be evaluated with leakier cell cultures or with intestinal tissue (animal or human).

The first method of choice for preliminary permeability experiments in our laboratory are thus CACO-2 cells and based on these results further necessary steps are undertaken to elucidate the absorption mechanisms pertaining to drug absorption.

2.2. CACO-2 Transporters and Enzyme Expression and the Comparison with the *In Vivo* Human Expression

Ethical issues make it impossible to obtain healthy human samples for high through-put studies so one has to settle for *in vitro* cancerous human or non-cancerous animal cells and tissues or other sub-cellular *in vitro* models to evaluate the contribution of the active transport (absorptive or secretory) to the drug's pre-systemic metabolism in the human intestine. Markedly different transporter and enzyme expression patterns in the *in vitro* models and *in vivo* models often impair the use of these tools to accurately evaluate the importance of transporter-enzyme interplay in drug pharmacokinetics. Therefore, a thorough analysis of gene expression has to be performed to obtain the insight into which proteins are expressed in sufficient levels and in the active form to qualitatively detect the involvement of transporter/enzymes in the studied processes.

Detailed studies by Volpe [9], Hayeshi [10], Maubon [11], Englund [12] and Sambuy [16] involving CACO-2 cells provided from American Type Cell Culture – ATCC and European Collection of Cell Cultures – ECACC sources, exposed to different culture conditions and protocols (i.e. seeding densities 63000 – 700000 cells/cm^2, passage number 26

- 98, media composition, cultivation time 15 or 21 days) have already been performed. The relative mRNA quantities as the measure of transporter/enzyme expression in CACO-2 cells were determined by real-time PCR [12, 13, 16]. Although transporter expression patterns were relatively similar between CACO-2 cells of different origins and of different passages, the study by Hayeshi et al. [10] pointed to high inter-laboratory variability of transporter expression resulting from different cultivation protocols, cell origin, passage number and time, cells have been cultivated. Namely, the coefficient of relative mRNA level variation was greater than 50%; i.e. the ratio between the highest and the lowest relative mRNA levels for some transporters ranged from 1 to 1000. Maubon et al [11] investigated transporter expression in CACO-2 cells from two sources (ECACC and ATCC) but within similar passage numbers (ranging from 30 to 40) and up to 2-fold quantitative differences in mRNA levels for the most expressed transporters were observed in both lines. Seithel [13] and Englund [12] compared the expression patterns between different segments of the small and large intestine (duodenum, jejunum, ileum, and colon) and sought the best correlation between CACO-2 cells transporter expression with individual intestinal segment. The authors established that the CACO-2 expression pattern simulated human jejunum best while the worst correlation was obtained between CACO-2 and colon. To illustrate the results of these studies, the summary of the main conclusions is presented in Table 1, which shows the rank order of transporter expression in CACO-2 cells and in different intestinal segments according to different authors. The expression patterns of the most important uptake and secretory transporters along human intestinal tract are shown in Figure 3, which was generated based on data published by Englund et al [12].

OATP-B expression in CACO-2 cells is the highest, followed by MRP-2, PepT1 and Pgp and other less expressed MRPs, BCRP, OCTs and OCTNs. OATP-B mRNA levels in CACO-2 are much higher than in human segments (up to 20-times higher expression) [12], whereas the expression of other transporters (i.e. MRP-2, Pgp, PepT1) is generally lower than *in vivo*. MRP-2 expression in CACO-2 cells is usually higher than Pgp, whereas for BCRP high controversity can be observed. Depending on the source of the cells, passage number and cell's cultivation time, BCRP expression has been reported to vary between moderate to low levels. However, in the human intestine, BCRP expression in all intestinal segments (duodenum, jejunum and ileum) is higher than the expression of the other two efflux pumps (i.e. MRP-2 and Pgp) – Figure 3. Pgp, MRP-2 and BCRP expression along human intestine increases from duodenum to jejunum, then slightly decreases in the ileum and is significantly lower in the colon (Figure 3A). PepT1 is the most prevalent uptake SLC transporter in all intestinal segments, followed by OATP-B and MCT-1, the last being the main transporter in colon (Figure 3B). Compared to CACO-2 SLC expression, only OATP-B is sufficiently expressed, while the expression of the other two is significantly lower.

Table 1. The review of studies by Volpe [9], Hayeshi [10], Maubon [11], Englund [12] and Sambuy [16] regarding transporter expression in CACO-2 cells (from American Type Cell Culture – ATCC or European Collection of Cell Cultures – ECACC) and in different segments of the human small intestine and colon

	Conclusion
Duodenum	Study by Maubon et al. [11]: BCRP ≅ PepT1 > OCTN-2 > Pgp ≅ MCT-1 ≅ MRP-2 > MRP-3 > OATP-B
Jejunum	Study by Hayeshi et al. [10]: highly expressed: Pgp, MRP-3, BCRP, PepT1; notable expression: MRP-2, OATP-B; not expressed: OCT-2, OAT-2 CACO-2 and jejunum qualitatively similar expression of Pgp, MRP-2, -3, OATP-B, PepT1 Quantitative differences: higher MRP-2, -3, 4, -5, -6, OATP-A, -B, OCT-1 and MCT-1 and lower CACO-2 lower BCRP and PepT1 expression in CACO-2 than in jejunum. Study by Englund et al. [12]: highly expressed: PepT1 >> MRP-2, ≅ Pgp; intermediate expression: BCRP > MRP-3 > OCTN-2 > MCT-1; low expression: OATP-B > OCT-1 > OAT-2 ≅ OCT-3 Quantative differences: CACO-2 lower Pgp (7x ↓), MRP-2 (2x ↓), PepT1 (10x ↓) expression. Intermediately expressed transporters: 2-7-times lower expression in CACO-2 than in jejunum. CACO-2 higher OCT-3 and OATP-B (13x ↑) expression than in human jejunum. Study by Maubon et al. [11]: BCRP > Pgp > PepT1 > OCTN-2 ≅ MRP-2 > OATP-B ≅ MCT-1 ≅ MRP-3
Ileum	Study by Maubon et al. [11]: BCRP ≅ Pgp ≅ PepT1 > OCTN-2 > MRP-2 ≅ OATP-B ≅ MCT-1 ≅ MRP-3 [12]
Colon	Study by Englund et al. [12]: highly expressed: MRP-3 >> Pgp; intermediate expression: MCT-1 ≅ OCTN-2; low expression: BCRP ≅ PepT1 > OATP-B ≅ OCT-3 > MRP-2 ≅ OCT-1 no correlation between CACO-2 and colon expression Study by Maubon et al. [11]: MCT-1 ≅ BCRP > OCTN-2 > Pgp > MRP-3 > OATP-B > PepT1
CACO-2	Study by Sambuy et al. [16]: high expression: OATP-B; moderate expression: MRP-2 > PepT1 > MRP-3 > MCT-1 > Pgp > BCRP > MRP-6; low expression: MRP 4, 5 > MRP-1 > OCT-1-3 Comparison of CACO-2 cells from different origins and passage numbers: transcellular permeability comparable, paracellular permeability significantly different Study by Hayeshi et al. [10]: high expression: Pgp, MRP-2, -3, OATP-A, -B, MCT-1, PepT1; low expression: MRP-1, -4, -5, -6, BCRP, OCT1, OCTN-1, -2 ECACC CACO-2: MRP-2 > MRP-3 > OATP-B > Pgp > PepT1 > OATP-A > MCT-1 ATCC CACO-2: MRP-3 > PepT1 > MRP-2 > OATP-B > MCT-1 > Pgp > OATP-A Study by Englund et al. [12]: intermediate expression: MRP-2 > OATP-B > PepT1 >> Pgp; low expression: MCT-1 ≅ MRP-3 ≅ BCRP ≅ OCTN-2 >> OCT-3 > OCT-1 > OAT-2; absent: OCT-2, OAT-1, -3, OATP-8 Study by Maubon et al. [11]: 4-day old CACO-2: MRP-2 ≅ OATP-B > MRP-3 > MCT-1 ≅ OCTN-2 ≅ Pgp ≅ PepT1 > BCRP >> OCT-1 21-day old CACO-2: OATP-B ≅ MRP-2 > Pgp ≅ PepT1 ≅ MRP-3 ≅ OCTN-2 > MCT-1 > BCRP >> OCT-1

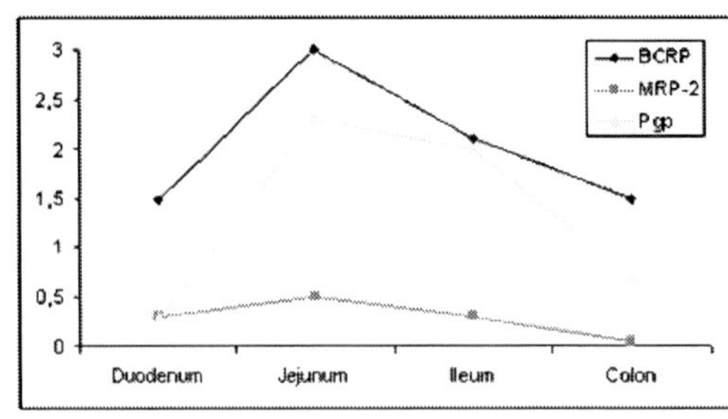

Figure 3A

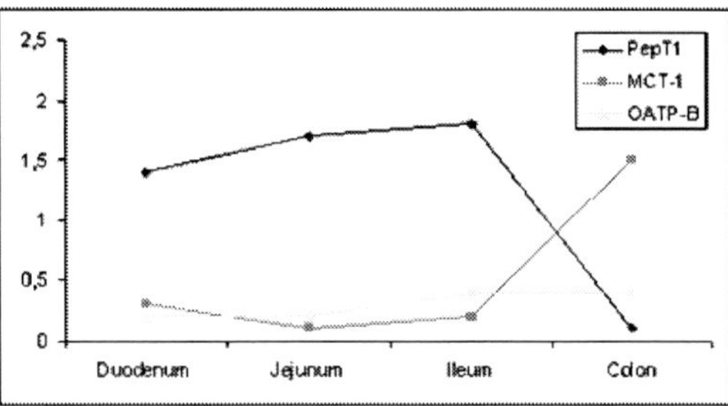

Figure 3B

Figure 3. the relative expression of efflux – ATP Binding Cassette Transporters (ABC transporters: BCRP, MRP-2, Pgp – Figure 3A) and uptake – Solute Carrier Transporters (SLC transporters: PepT1, MCT-1, OATP-B – Figure 3B) normalized to the expression of villin along human intestine. Figures are based on the data from Seithel et al. [13].

CACO-2 cells also lack the expression of pivotal metabolizing enzymes, although the cell genome contains genes coding for these proteins. According to Seithel [13], glutathione-S-transferase 1A is highly expressed, whereas only moderate levels of CYP2C19, CYP2C9 and sulfotransferase 2A1 have been detected. Other studies reported on sufficient levels of hydrolases (sucrase-isomaltase, lactase, amino-peptidase N, dipeptidylpeptidase IV, alkaline peptidase), carboxylesterase, CYP1A1, 1A2 [29] and of phase II enzymes (glucuronosyltransferase, N-acetyltransferase, sulfotransferase) [9]. However, it is generally accepted that unless CACO-2 cells could be exposed to special cultivation conditions (by adding 1, 25-dihydroxy vitamin D_3 or drugs with known inductive potential) or genetic manipulation (i.e. transfection) to induce the genes for metabolizing enzymes, the predictive power of CACO-2 cells in assessing the participation of enzymes in the intestinal first-pass metabolism is compromised.

On the other hand, the complex interplay between transporters and enzymes (Pgp and CYP3A4 or MRP-2 and UGT) often precludes one to accurately determine the impact of individual protein - either transporter or enzyme - in a drug′s absorption. Therefore in these situations, where such information is needed, the use of less complex *in vitro* CACO-2 models enables unequivocal evaluation of the transporter`s impact on the absorption and the metabolism could be assessed afterwards with less/more complex models (i.e. human microsomes, *in vitro* animal/human models).

Although CACO-2 cells do express individual transporters and enzymes to an appreciable extent (Table 1), this does not necessary lead to sufficient protein activity [30]. The correlation between Pgp and PepT1 mRNA quantities and Pgp and PepT1 mediated active transport has been observed [12] but for other proteins further studies are warranted to establish the relationship between mRNA levels, protein levels and transporter functionality. Namely *in vivo* transporter function is often influenced by constantly changing intraluminal conditions along the GI tract (microclimate pH, ions gradients, nutritional supply, and digestive enzymes) that could not always be simulated with *in vitro* CACO-2 models.

Based on the provided review on transporter and enzyme expression in CACO-2, the involvement of highly expressed uptake (OATP-B) and secretory (Pgp, MRP-2) transporters in the drug absorption from GI tract could easily be evaluated. In other cases, the participation of intermediately expressed transporters in drug transport is difficult to assess unless very sensitive analytical methods could be applied or if the drug exerts very high affinity for an individual transporter. Otherwise, one has to either transfect CACO-2 cells with the transporter gene, induce its expression or eventually replace CACO-2 cells with alternative *in vitro* models.

3. CACO-2 Cells as an *In Vitro* Prognostic Tool for Assessing Pharmacokinetic Drug-Garlic Interactions

The consumption of herbal supplements is witnessing a meteoric rise with sales running into billions of dollars yearly [31]. Due to their natural origin, the widely held consumers' view of low risk for deleterious interactions between herbal supplements and conventional prescription medications is unfortunately misleading. In spite of the *in vivo* and *in vitro* data indicating significant pharmacodynamic and pharmacokinetic interactions, increasing physicians' or patients' awareness of widespread adverse effects still seems an ominous task [32 - 34]. In attempt to aid and improve their health status, the elderly and chronically ill, the main subpopulation of supplement users [35], also often do not disclose their use of complementary and alternative methods and products with their physicians. Simultaneously, these poly-pharmaceutical patients are more prone to interactions, because the functional hepatic, intestinal and kidney reserves decline with aging [36 - 38].

Preventive and curative garlic pharmacological properties (i.e. antimicrobial, hypoglicaemic, lipid-lowering, immunomodulatory, antihypertensive, antitumorogenic) [39, 40] make garlic supplements an attractive alternative to cardiovascular and chronic patients world-wide, suggesting a high possibility of co-administration with synthetic drugs and therefore high incidence of synergistic pharmacodynamic effects [41 - 43] or altered expression and/or activities of transporters and enzymes [44 - 49] involved in drug's absorption, distribution, metabolism and excretion.

Predicting toxic and therapeutic drug profiles with competent, simple, robust and reproducible *in vitro* models as an alternative to clinical studies has become an industrial and regulatory priority. Although *in vitro* and *in vivo* animal (mouse, rat, dog, monkey) or cellular biotransformation and absorption models for detecting drug-drug and drug-phytochemical interactions are readily available, their simplistic and static nature limits the predictive power of such experimental approaches to correctly forecast the *in vivo* drug plasma profiles. The *in vitro* ascertained pharmacokinetic interactions usually demand an additional conformation of the results with the *in vivo* animal studies, whereas in the reverse situation (i.e. the absence of *in vitro* interactions) there is no need for further *in vivo* studies [1, 50, 51].

The existing *in vivo* data indicates significant influence of garlic co-administration on the extent and rate of human immunodeficiency virus protease inhibitor (HIV-PI) saquinavir absorption, which consequently affects a drug's efficiency and safety [49, 52]. No studies that have been performed elucidate the mechanisms responsible for the observed garlic – HIV-PIs interactions. Due to the wide variety of putative garlic medicinal uses, there is a strong possibility that the absorption of other drugs, prescribed for the treatment of cardiovascular, diabetic and antiviral conditions could also be modified. Therefore the applicability of CACO-2 cell monolayers to qualitatively detect garlic-drug pharmacokinetic interactions has been investigated by evaluating the impact of garlic supplement aged garlic extract (AGE) on the activities of uptake and efflux transporters in CACO-2 and compared to the effect AGE exerted on transporters expressed in rat small intestine. The permeabilities of drugs – substrates for the transporters (Table 2) – were monitored through CACO-2 cell monolayers and compared to the results obtained with rat small intestine *in vitro*.

Before the evaluation of garlic's impact on transporters, the permeabilities (P_{app}) of all selected model substrates were determined in mucosal-to-serosal/apical-to-basolateral and in the opposite direction, serosal-to-mucosal/basolateral-to-apical to confirm the participation of investigated transporter in the transport of selected drug. The involvement of a transporter in drug's transport leads to higher permeability in one direction; in the case of efflux transporter the permeability is higher in serosal-to-mucosal/basolateral-to-apical direction, whereas in the case of uptake transporter, the converse is true and the permeability in mucosal-to-serosal/apical-to-basolateral direction exceeds the permeability in the other. If the addition of a transporter inhibitor (PSC833, verapamil for Pgp, MK571, benzbromarone for MRPs, fumitromorgin C for BCRP, amoxicillin for PepT1, DIDS or CHC for OATPs and MCTs) lowers the permeability in the direction of the active transport, this is reliable evidence for the involvement of specific transporters in drug transport.

The asymmetrical permeability properties were observed for all stated substrates through CACO-2 monolayers and rat intestine (Table 2), indicating adequate transporter expression and activity. The drug selection for the investigation of garlic influence on individual transporter activities was thus appropriate [53, 54].

In the presence of AGE (at luminal concentration of 1 v/v%; the recommended human daily dose calculated to 250 ml) significant transporter activity changes were observed. As presented in Table 2, Pgp, MRP-2, OATP and MCT-1 mediated drug transport increased in both *in vitro* models, whereas BCRP and PepT1 transporters responded differently to the presence of garlic phytochemicals depending on the model used.

The activities of BCRP and PepT1 decreased in CACO-2 cells, while an increased transport was noted in rat small intestine. The underlying reasons for these discrepancies in transporter responsiveness to garlic phytochemicals were explained by differences in transporter expression, low amino acid homology and different transporter regulation pathways in cancerous versus non-cancerous cells [54, 55].

Up to 100-times lower BCRP expression in CACO-2 cells than in human jejunum has been detected [56, Table 1]. Lower BCRP apical expression ensures significantly lower BCRP-mediated transport. Therefore, intracellular retention of BCRP substrates increases, which is especially important for cancerous cells. Namely, BCRP substrate protoporphyrine IX is an important regulator of activities of proteases involved in cancerous physiology.

These enzymes participate in cell division and enable constant replication [57]. Such processes are absent in non-cancerous cells. In spite of these differences, CACO-2 cells with sufficient expression of BCRP (CACO-2 cells with high passage number, transfected CACO-2 lines) have proven useful in identifying BCRP substrates/inhibitors.

Table 2. The impact of aged garlic extract (AGE) on the activity and transports of the corresponding substrates (cardiovascular, hypoglicaemic and antiviral drugs), determined with CACO-2 cell monolayers and rat small intestine [53, 54]

Transporter	Substrate	Experiment al design	R_{CACO-2}	R_{rat}
Pgp	Digoxin Hydrochlorothiazide Glibenclamide Darunavir	A	↑↑	↑
MRP-2	DNP-SG	A	↑	↑
Pgp + MRP-2	Losartan Valsartan Furosemide Atorvastatin Pravastatin Saquinavir	A	↑	↑
BCRP	Rosuvastatin	A	↓	↑
PepT1	Valacyclovir	B	↓	↑
OATP + MCT-1	Atorvastatin Pravastatin	B	↑	↑
OCT1	Ciprofloxacin	A	=	=

Experimental design A: the permeability of drug was monitored in serosal-to-mucosal (S-M)/basolateral-to-apical (BL-AP) direction through rat small intestine and CACO-2 cell monolayers, respectively. The pH of incubation solution on both sides of tissue/cells was adjusted to 7.4 (Ringer buffer). R_{CACO-2}, and R_{rat} represent the ratio between S-M/BL-AP permeabilities after the addition of AGE and the corresponding S-M/BL-AP permeabilities determined before AGE addition.

Experimental design B: the permeability of drug was monitored in mucosal-to-serosal (M-S)/apical-to-basolateral (AP-BL) direction through rat small intestine and CACO-2 cell monolayers, respectively. The pH of incubation solution on apical/mucosal side was adjusted to 6.5, whereas pH was on the serosal/basolateral side was set to 7.4.

↓ - a significant decrease of the R values after AGE addition, indicating a decrease of the transporter's activity, and thus a decrease of the transport of the corresponding substrate.

↑ - a significant increase of the R values after AGE addition, indicating an increase of the transporter's activity and thus an increase of the transport of the corresponding substrate.

= - unchanged R values after AGE addition, indicating to impact of AGE on transporter's activity and thus on the transport of the corresponding substrate.

This means that BCRP substrate specificity in cancerous cells is similar to the BCRP specificity in human intestine and the opposite permeability changes observed for rosuvastatin after the addition of AGE could only be caused by the dissimilarities in BCRP regulation in cancerous cells, which sustains uncontrolled cell division, but is absent in normal tissue.

PepT1 also responded differently to the presence of AGE in CACO-2 cells than in rat jejunum. Masereel et al [58] established that the CACO-2 PepT1 amino acid sequence showed no homology with human PepT1 variants, whereas in rodents, PepT1 transporters' homology with human PepT1 was greater than 80%. Besides PepT1 structural dissemblance we also observed that CACO-2 PepT1 transport activity failed to increase in the presence of flavonoids, which are known PepT1 activators [59]. Because flavonoids are also present in AGE and CACO-2 expressed PepT1 failed to respond to their presence, one is precluded in assessing the physiologically relevant AGE influence on PepT1 transport through CACO-2 cell monolayers. It is thus necessary to evaluate PepT1-garlic interactions with rat small intestine or other cell lines, whose amino acid sequence homology with human PepT1 is higher.

Both models therefore identically forecasted the impact of garlic phytochemicals on the activities of ABC transporters Pgp and MRP-2 and of SLC transporters OATPs, MCT-1 and OCT-1, whereas we were unable to study BCRP- or PepT1-garlic interactions. The increased Pgp and MRP-2 activities and consequently increased substrate transport were caused by positive cooperation effect [60]. Both secretory proteins enable simultaneous binding of multiple substrates, because there are four binding sites in Pgp and two in MRP-2 [60, 61]. The binding of one substance (i.e. herbal constituents) to individual Pgp or MRP-2 binding site induces structural changes in the protein, which affects the structure and consequently the binding affinity of the remained unoccupied binding sites. Their binding affinity increases, resulting in higher transport of other MPR-2/Pgp substrates as observed in the case of garlic phytochemicals and investigated drugs. It is also possible that both compounds (garlic phytochemicals and drug) would bind to the same binding site. This could happen, if the concentration of garlic extract could be so high that the phytochemical would completely occupy its binding site and afterwards it would non-specifically bind also to the other, drug–binding Pgp/MRP-2 site. It is also possible that both substances would compete for identical binding site from the start. In the latter case, one could observe a competitive inhibition of drug transport in the presence of garlic phytochemicals, but this was not the case in our study.

Facilitated diffusion by MCT-1 and OATPs in the presence of AGE also significantly increased. These co-transporters do not have multiple binding sites for their substrates but enable drug transport as a co-transport with ion (Na^+, H^+, cation) [62]. In the presence of AGE, significant electro-physiological changes (potential difference, short-circuit current) in CACO-2 cells and rat small intestine were observed, indicating alterations in the ionic buffer composition on the mucosal/apical side of both *in vitro* models mounted in side-by-side diffusion chambers [59]. Garlic extract contains pungent herbal constituents and flavonoids, which activate apical electrogenic calcium (TRP-V) and chloride (CFTR) channels. This led to an increased calcium absorption and chloride secretion, altering ionic equilibrium and affecting facilitated diffusion [53, 54].

Whether *in vitro* observed altered transporter activities could significantly impact the fraction of drugs absorbed *in vivo*, depends on the drug properties (lipophylicity, ionization, size, charge…), the expression of that transporter along the GI tract and the properties of the administered formulation. The selection of pharmaceutical drug delivery systems to manipulate the delivery of compounds to their absorption windows in order to avoid pre-systemic drug metabolism, the impact of transporters, drug degradation or to prolong drug release, involves either the use of immediate or modified release dosage forms. Immediate release formulations allow fast dissolution of at least 85% of the administered BCS I or acidic BCS II therapeutic doses [17], giving sufficiently high intraluminal drug concentrations, which saturate efflux, uptake transporters or enzymes. Therefore in these cases garlic co-administration will not pose any danger in achieving therapeutic plasma drug concentrations. However if modified release formulation was prepared with the same drugs, then the rate-limiting, slower drug release kinetics from the formulation would give lower intraluminal drug concentrations. In such situations, neither transporter nor enzymes would be saturated and phytochemical-drug pharmacokinetic interactions could be anticipated to affect F_{abs}. Independently of the selection of formulation (immediate or modified release), herbal phytochemicals are believed to affect the absorption of all other BCS II, III or IV classified compounds, if these phytochemicals can affect the activity or expression of transporter/enzymes participating in drug absorption [64, 65].

Changed F_{abs} values are usually considered to affect plasma drug levels. However, if the drug plasma concentrations remain inside the therapeutic window, there are no clinically relevant efficiency or safety issues. Such assumptions would be unjustifiable in the case of drugs with narrow therapeutic windows (digoxin, HIV-PI, warfarin), where small changes of transporter/enzyme activities in the intestine or liver could be intimately associated with too low or too high plasma concentrations, responsible for undesirable adverse effects, toxicity or therapeutic failure. This is supported by the *in vivo* study, where digoxin and quercetin (flavonoid, ubiquitous phytochemical) were co-administrated and as a result a sudden death in two out of three pigs were noted. The deaths were caused by significantly elevated extent (170% increase) and rate (410% increase) of digoxin absorption due to Pgp inhibition by quercetin [66], which elevated digoxin plasma concentration above maximal plasma levels. Similarly, therapeutic drug monitoring is also recommended for HIV-PIs (saquinavir, darunavir) [67], because plasma drug levels below the minimal therapeutic plasma concentrations lead to viral mutations, drug resistance and therapeutic failure. In situations, when the attained plasma HIV-PIs concentrations would exceed maximal therapeutic plasma levels, adverse drug reactions (especially GI irritations) could diminish patient's compliance and in the long-run, too high HIV-PI plasma levels would lead to metabolic syndrome [68, 69]. Treating HIV infected patients for metabolic syndrome further obligates the patient to an even greater pill burden and higher incidence of pharmacokinetic interactions. Therefore, it is imperative to assure patient's compliance by informing him/her of the potential deleterious GI symptoms during undisclosed self-medication with herbs.

According to Table 2, Pgp/MRP-2 mediated digoxin, saquinavir and darunavir secretion from cells significantly increased in the presence of AGE. Based on the literature data above [66, 67], self-medication with garlic supplements and digoxin, saquinavir or darunavir should be avoided due to higher probability of lowering F_{abs} and therapeutic failure. For other tested compounds (Table 2), F_{abs} could change, but most probably would not pose serious threat for the therapeutic outcome because neither of them is classified as a drug with a narrow therapeutic window. Besides, some compounds, i.e. statins (atorvastatin, pravastatin) are substrates for uptake and efflux transporters and increased activities of both types of transporters affect plasma drug levels in the opposite manner; increased uptake transporter activities contribute to higher plasma levels, whereas increased efflux transporter activities tend to lower plasma levels. Therefore for such promiscuous compounds, the final influence of garlic supplements depends on the process (uptake or efflux) with faster kinetics and higher capacity. Furthermore, the presented phytochemical interactions could also curtail the hepatic distribution and metabolism by either modifying the activities of enzymes and of sinusoidal and canalicular ABC and SLC transporters or by modifying the protein expression during long-term concomitant consumption. These hypotheses have already received some experimental support [70, 71] and will undoubtedly remain at the forefront of first-pass metabolism research and drug-phytochemical interactions for several years.

4. CACO-2 CELLS FOR THE EVALUATION OF SELF-MICROEMULSIFYING DRUG DELIVERY SYSTEMS (SMEDDS)

4.1. Self-Microemulsifying Drug Delivery Systems:The Objectives and Advantages

Due to the ease and full patient's control of administration, the oral route remains the most often considered method for administration of drugs. Nevertheless, in recent years, the number of drugs that present challenges to the development of products with sufficient oral bioavailability (classes II-IV according to BCS) is increasing. Many of those new chemical entities fail to proceed to the advanced stages of research and development in spite of their pharmacological activity. Solubilization of hydrophobic drugs with low aqueous solubility has been a major area of interest in recent years. The most commonly encountered means which can be utilized to turn these molecules into orally administrated medications with sufficient bioavailability include molecular changes as well as formulation strategies. Drug dissolution and solubilization can be enhanced via the synthesis of prodrugs, particle size reduction, formulation of solid dispersions, co-precipitatates, complexation with cyclodextrins, as well as by drug incorporation into lipid-based systems [72 - 75].

Lipids and lipophilic excipients can have significant and beneficial effects on the absorption and exposure of co-administered drugs [76]. Formulations based on those excipients are called lipid-based systems, and range from simple oil solutions to complex mixtures of oils, surfactants, and hydrophilic co-solvents. According to the Lipid formulation classification system [77, 78] those systems are divided into four formulation classes (Type I-IV) based on their composition and ability to prevent drug precipitation during dispersion and digestion of formulation following oral administration (Table 3). In recent years much attention has been focused on Type III formulations that are further divided into two subclasses to distinguish between systems with higher lipid content (Type IIIa; self-emulsifying systems - SES) and more hydrophilic systems (Type IIIb; self-microemulsifying systems - SMES). Both types of formulations are isotropic mixtures of oil, a surfactant, one or more hydrophilic co-solvents or co-surfactants, and the drug. Upon dilution in aqueous media (i.e. GI fluids), followed by agitation provided by digestive motility of the stomach and the intestine, this system forms fine oil-in-water (o/w) emulsions or microemulsions [79, 80]. Type IIIa systems typically produce emulsions with a droplet size between 100 and 300 nm, while formulations that disperse to form transparent colloidal systems with droplet size less than 50 nm are usually referred to as SMES [81].

Table 3. Classification system and typical properties of lipidic formulation as proposed by Pouton et al [77, 78]

System	Typical composition	Characteristics	Advantages	Disadvantages
Type I	Glycerides (MG, DG, TG) (100%)	Non-dispersing, requires digestion	GRAS status; simple; good capsule compatibility	Formulation has poor solvent capacity unless drug is highly lipiphilic
Type II	Glycerides (MG, DG, TG) (40-80%) and water-insoluble surfactants (20-60%)	SEDDS formed without water-soluble components	Unlikely to lose solvent capacity on dispersion	Turbid o/w dispersion (droplet size 0,25-2 μm)
Type IIIA	Glycerides (MG, DG, TG) (40-80%), water-soluble surfactants (20-40%), and hydrofilic co-solvents (0-40%)	SEDDS formed with water-soluble components	Clear (IIIb) or almost clear (IIIa) dispersion; drug absorption without digestion	Possible loss of solvent capacity on dispersion; less easily digested
Type IIIB	Gliceridi (MG, DG, TG) (<20%), PAS s HLB>12 (20-50%), hydrophilic cosolvents (20-50%)	SMEDDS formed with water-soluble components.		Higher possibility for precipitation of drug on dispersion
Type IV	PAS s HLB<12 (0-20%), PAS s HLB>12 (20-80%), hydrophilic cosolvents (0-80%)	Furmulation disperses typically to form a micellar solution	Formulation has good solvent capacity for many drugs	Likely loss of solvent capacity on dispersion; may not be digestable

The optimum concentration of oil, surfactant and co-surfactant necessary to promote self-emulsification are determined by construction of a pseudo-ternary phase diagram, which should also assess the effect of drug loading on efficiency of self-emulsification. Development and characterization of SES and SMES has been extensively reviewed [81 - 84].

The number of excipients, which are pharmaceutically acceptable in the formulation of lipid-based systems, is large. The most popular ones used in lipid-based products for oral administration were reviewed by Gibson et al. [76], whereas some regulatory considerations in the use of those excipients for pharmaceutical development were recently presented by Mei-Ling et.al [85].

SMES typically contains long- or medium-chain natural or semi-synthetic triglycerides, or correspondingly more polar mono-/di-glyceride lipid blends. Hydrophilic non-ionic surfactants with an HLB value of approximately 12 or greater are most commonly used among surfactants. Bulky surfactants such as polysorbates or polyethoxylated vegetable oils are less toxic than single-chain surfactants and esters, which are also less toxic than non-digestible ethers. According to literature, semi-synthetic polyethylene glycol (PEG) derivatives of glycerides and fatty acids, polyglyceryl fatty acid esters, phospholipids, the sorbitan esters and their ethoxylated derivatives (polisorbates) are the most popular.

Typically, hydrophilic co-solvents (PEG 400, propylene glycol, ethanol and glycerol) are also used in SMES in order to facilitate the dissolution of large quantities of either the hydrophilic surfactant or the drug in the lipid base, and to aid dispersion of systems [80, 81, 83, 85].

Beside regulatory issues (i.e. irritancy and toxicity), solubilization capacity, miscibility, self-dispersibility, digestability, and fate of digested product as well as chemical stability of excipients has to be considered when SMES are formulated [83]. When developing SMES one should also be aware that the process of self-emulsification is specific to the nature of the oil/surfactant pair, the surfactant concentration, oil/surfactant ratio, and the emulsification temperature [81]. Furthermore, a number of typically employed excipients are susceptible to enzymatic degradation in the GI tract, therefore the absorption of lipophylic drugs from lipid-based formulations is determined not only by the patterns of dispersion but also by the digestion and solubilization of formulation-derived lipid and a co-administered drug in the GI tract [86]. Namely, the processing of exogenous lipids begins in the stomach, which is also the primary site of initial lipid emulsification in the GI tract. Triglycerides are hydrolized by the acid-stable lipases to di-glicerides and fatty acids. These products, along with the shear movement of the stomach and the passage through the pyloric sphincter into the duodenum, cause the formation of a crude emulsion. The presence of exogenous lipids in the duodenum stimulates the secretion of bile containing bile salts, phospholipids and cholesterol, and pancreatic fluids containing pancreatic lipase/co-lipase. These agents adsorb to the oil/water interphase and produce more stabilized emulsion with reduced droplet size. The enzymatic degradation of triglycerides in the small intestine is completed by the action of pancreatic lipase/co-lipase complex resulting in corresponding 2-monoglicerides, and two fatty acids. Those amphiphilic lipid digestion products together with bile lipids further promote dispersion of lipid droplets and generation of a series of colloidal species including vesicles and mixed micelles. By partitioning the co-administered drug substance into those colloidal structures, solubilization of poorly water soluble drugs can be enhanced, leading to reduced precipitation in aqueous GI milieu. Following uptake into the enterocytes, lipid digestion products with co-administered drug are exposed to the biochemical barriers, e.g. enterocyte cytochrome P-4503A4 enzymes, and efflux transporters that affect the magnitude of their transfer into the lamina propria. Once inside the cell, drug and lipid components can be transported to the systemic circulation either by portal vain or by the intestinal lymphatic system. Further details of lipid digestion, absorption and metabolism are available in the following reviews [76, 83, 87 - 89].

For drug to be well absorbed from the GI tract and provide good oral exposure, a number of potentially limiting factors must be overcome. Poor drug intraluminal stability, solubility and/or intestinal permeability, extensive enzymatic degradation and first-pass metabolism are the main objectives that usually render per oral administration of nearly half of the marketed drugs quite ineffective [90, 91]. Oily droplets of colloidal size that are spontaneously formed upon dilution of SMES with intestinal fluids advantageously deliver the drug in its dissolved form throughout its transit in the GI tract, thus providing a large interfacial surface area for drug absorption, and possibly avoiding the drug dissolution step [76, 77]. Furthermore, lipids delay gastric emptying and prolong formulation retention in the absorption window (i.e. stomach, small intestine) giving the drug an extended period of time to dissolve in the fluids and permeate the intestinal mucosa. However, lower droplet size is not always better as today we are aware that the performance of lipid-based delivery systems is governed by their fate in the GI tract, rather than the droplet size of the initial dispersion [83].

Apart from solubilization, the choice of an appropriate formulation can contribute to transporter/enzyme inhibition and increased GI membrane permeability. Furthermore, SMES containing long-chain lipids may also stimulate production of chylomicrons, and thus enhance lymphatic transport of lipophilic drugs directly to the systemic circulation, avoiding hepatic first-pass metabolism [76, 77, 88, 92, 93]. The choice of surfactants and lipids is therefore a crucial factor for the *in vivo* fate of the formulation and co-administered drug. When screening SMES for optimal composition, physicochemical properties of the drug substance and the interaction of the formulation with the GI tract, have to be considered carefully.

4.2. Studies Performed to Evaluate SMEDDS Impact on Intestinal Absorption of Drugs

The mechanisms underlying the abovementioned physiological changes leading to higher drug absorption have been extensively studied with different animal and cellular models [94, 95]. The employments of *in vivo* cannulated (un)anesthetized rats, mice (wild type or genetically modified) and dogs has helped in determining the amount of drug absorbed into the intestinal lymphatic system. Extremely lipophilic (logP > 5, the solubility in lipids > 50 mg/ml) and large drugs are prevented from transversing intestinal mucosa and from entering the portal blood supply, but can easily pass into the more open lymphatic capillaries of central lacteal in the villus, whose content is reunited with systemic blood supply at the junction of the left jugular and the left subclavian veins, avoiding hepatic pre-systemic metabolism. Because ingested fat (especially oleic acid) stimulates intestinal production of chylomicrons, LDL and VLDL particles, the transport of drugs incorporated into these structures by lymph also increases. Besides that, some of the lipophilic SMES constituents increase paracellular and transcellular permeability, inhibit Pgp and CYP3A4, which has been proven with *in vitro* cellular/tissue models. CACO-2 cells are the most widely used due to their capability to simulate certain steps of lymphatic uptake of drugs [91]. The use of CACO-2 cells has also improved the knowledge regarding the mechanisms behind the permeability increase. Namely some surfactants in lipid vehicles can affect the architecture of tight junction proteins (actin filaments, zona occludens 1), open the pores and in this way increase the paracellular diffusion [96, 97]. On the other hand, oil droplets of microemulsion tend to have high surface area and come into close proximity with the absorptive intestinal cells, which expedites the distribution of amphiphilic substances (solubilization agents, enhancers of intestinal penetration) and drugs from microemulsions into the membrane. Surfactants tend to increase membrane fluidity, which results in increased transcellular permeability and inhibition of efflux proteins (Pgp, MRP-2) [95]. Therefore more drug permeates the apical membrane, entering either into the blood or lymph supply and at the same time, less of the drug is subjected to CYPs metabolism and efflux from the absorptive cells back into the GI lumen. Intracellular drug concentration increases, causing a higher concentration gradient between the cells and the circulation, further aiding the absorption.

Experimental data investigating drug absorption in the presence of different lipid vehicles has been obtained with the abovementioned animal and cellular models for pacitaxel, amprenavir, saquinavir, ritonavir, digoxin, celiprolol, amphotericin B, cyclosporine A, halofantrine and has aided in the registration process of cyclosporine A, pacitaxel, saquinavir and ritonavir formulated into the lipid based pharmaceutical dosage forms [98].

4.3. The Mechanism Responsible for Increased Absorption of Furosemide Incorporated in SMEDDS

To demonstrate advantages of SMEDDS in the per oral absorption of drugs with unfavorable physical, chemical or biological properties, a low permeable and low soluble BCS IV class drug, furosemide, was loaded into the Ca-pectinate microcapsules with self-microemulsifying cores, which merge the advantages of SMES with those of a solid dosage forms. Afterwards the mechanisms enhancing furosemide's solubility, dissolution and permeability were investigated [74].

The microcapsules were formulated by liquid jet co-extrusion with Inotech IE-50R encapsulator equipped with a concentric nozzle. The shell forming phase (pectin solution) envelopes the core phase (furosemide-loaded SMES) as they flow together through the concentric nozzle, forming a continuous jet that is broken apart into droplets by vibration of the membrane. When these droplets come in contact with the hardening solution, a sol–gel transition of pectin in the shell occurs, encapsulating SMES with the drug in the core [74].

Firstly, the liquid core phase was optimized to obtain stable and homogeneous mixture of medium chain triglyceride Mygliol 812® (lipophylic phase), caprylocaproyl macrogolglycerides Labrasol® (surfactant) and polyglyceryl-6 dioleate Plurol Oleique® (cosurfactant), which ensured the highest furosemide solubilization capacity (ca 86 mg/ml, which is 212-fold higher than furosemide's water solubility) and at the same time the largest possible surface area formed upon *in vivo* microemulsion dilution. Furosemide and 0.5 mg $CaCl_2$/g SMES were than added to the SMES that was allowed to blend for 24-hours. Finaly, furosemide-loaded SMES was thickened by 4% white wax or colloidal silica. The main challenge in production of microcapsules with SMES in the core is to prevent the undesired mixing of the self-microemulsifying core with the aqueous shell forming phase, which results in core leakage during the microcapsules production process (i.e. during hardening of microcapsules in 0.5 M $CaCl_2$ solution, and during their drying in the fluid bed system). The core retention and consequently encapsulation efficiency was considerably improved by thickening SMES with colloidal silica or white wax, together with the addition of the $CaCl_2$ that promoted cross-linking of the polymer in the shell phase from the inside out, thereby limiting the undesired mixing between the core and shell forming phase. To reinforce the microcapsules' shell, chitosan coating was additionally applied and the microcapsules were then rinsed and dried in a fluid based system. Afterwards, the release of furosemide from microcapsules with a self-microemulsifying core was compared to drug release from reference microspheres (without SMES in the core) that were prepared from the shell phase loaded with 1 % of furosemide, following the same procedure as microcapsules. *In vitro* dissolution tests clearly indicated significantly improved furosemide release profile (Figure 4). The dissolution rate of furosemide proved to be considerably faster from microcapsules, where furosemide was solubilized in the self-microemulsifying core (55% of furosemide released during 4 hours in 900 ml of buffer pH 3, USP apparatus 2 at 37°C, 75 rpm) than from reference microspheres (only 5% furosemide release from reference microspheres in the same time-frame). Obtained results are in agreement with the ability of SMES to avoid the drug dissolution step in the GI tract by delivering drugs in dissolved form. When the drug is administered in a form of multiparticulate dosage forms, e.g. microcapsules, disintegration of formulation can also be avoided to a certain degree. Namely, the officinal peroral dosage forms (i.e. tablets, capsules) have to disintegrate into smaller particles before drugs release and dissolution can begin, whereas the SMES delivery system investigated in this study would have readily disintegrated during the process of the stomach emptying into the duodenum, while mild agitation and intestinal secretions would enable the formation of microemulsions with solubilized furosemide [74].

The impact of SMEDDS constituents on intestinal furosemide permeability was then investigated with CACO-2 monolayers and rat small intestine (duodenum, jejunum, and ileum) mounted in side-by-side diffusion chambers *in vitro*. Reference permeability experiments with water furosemide solution revealed that this drug displayed very low permeability in the absorptive direction through CACO-2 cell monolayers and through all segments of rat small intestine. In CACO-2 cells apical-to-basolateral permeability was below 1×10^{-6} cm/s and in all segments of rat's small intestine mucosal-to-serosal permeability was lower than 1×10^{-5} cm/s. The absorbed furosemide quantities were further diminished by profound efflux of furosemide by MRP-2 and partially also by Pgp transporters from cells back into the mucosal/apical solution [74].

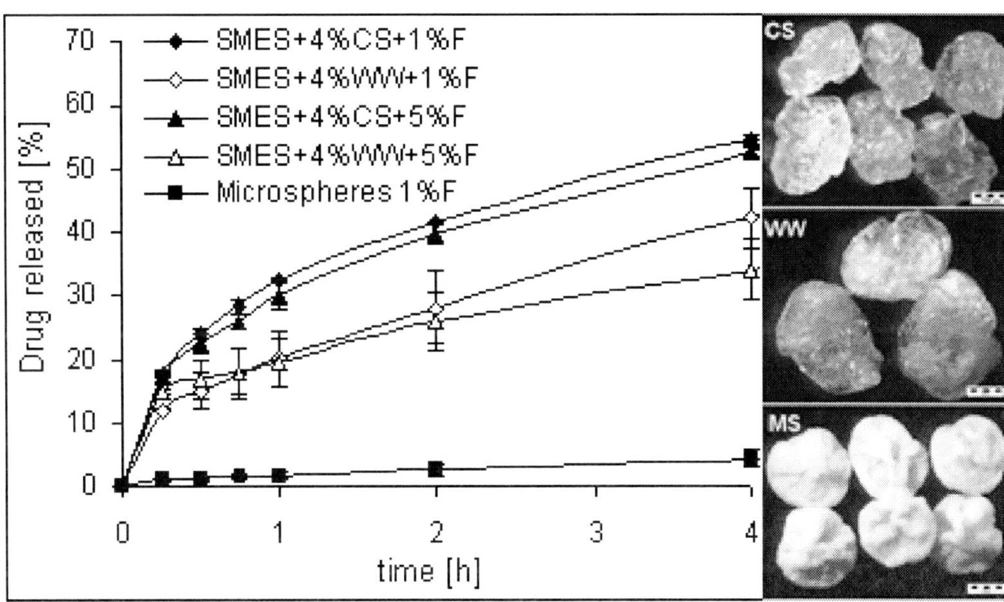

Figure 4. Percentage of furosemide (F) released from microcapsules with SMES in the core (thickened with colloidal silica (CS) or white wax (WW) and containing 1 or 5 % of furosemide) or reference microspheres (without SMES in the core loaded with 1 % furosemide). *In vitro* release experiments were conducted in hydrochloric acid solution, pH 3; n=3.

Applying furosemide in SMEDDS to apical/mucosal side of CACO-2 cell monolayers or rat small intestine, respectively, resulted in significantly increased absorptive permeabilities

(Figure 5) and in significant inhibition of Pgp and MRP-2 furosemide transport (basolateral-to-apical CACO-2 furosemide permeability in the presence of SMEDDS decreased from $(1.4 \pm 0.0)\times10^{-6}$ cm/s to $(1.0 \pm 0.1)\times10^{-6}$ cm/s; serosal-to-mucosal furosemide permeability in duodenum decreased from $(16.7 \pm 3.9)\times10^{-6}$ cm/s to $(5.1 \pm 0.5)\times10^{-6}$ cm/s in the presence of SMEDDS).

Increased absorptive furosemide permeabilities detected with both *in vitro* models resulted from close contact between the apical/mucosal membrane of the absorptive cells and oil droplets. This enhanced rapid redistribution of surfactants from microemulsions into the lipid bilayer, created polar membrane defects responsible for changed membrane fluidity [74, 75], favoring absorption of less permeable compounds. Additionally, we also observed increased permeability through tight junctions in the presence of SMEDDS by monitoring the permeability of paracellular marker fluorescein. Lipophilic SMEDDS components (i.e. Labrasol) have been shown to irreversibly modulate tight junction architecture when given in sufficiently high concentrations [74]. Because an irreversible tight junction opening usually leads to tissue/cell injury, such high Labrasol concentrations were not used in the formulation of our SMES. Instead of 1% w/v of Labrasol, which is known to lower CACO-2 cell monolayer integritiy and thus viability [74], in our experiments Labrasol was present at 0.2 % w/v (the concentration which would be attained if all Labrasol from SMEDDS would incorporate into the lipid bilayer). This concentration was low enough not to hamper *in vitro* models (transepithelial electrical resistance (TEER) values did not change after SMEDDS addition), but was sufficient enough to cause a moderate increase of fluorescein paracellular permeability. Our findings may therefore lend a support to the hypothesis that increased absorptive permeability of investigated BCS IV model drugs incorporated in SMEDDS results from increased transcellular and partially also by paracellular permeability. Additionaly, components of lipid vehicles also inhibited intestinal efflux transporters Pgp and MRP-2, by incorporating them into the lipid bilayer, affecting membrane fluidity, and consequently the activities of transmembrane proteins [75] or they could go directly to Pgp and MRP-2 transporters [99], displacing furosemide and causing less of the drug to be effluxed back into the GI lumen.

Based on the presented results, CACO-2 cells were found to be a valuable tool in assessing the impact of SMEDDS on the intestinal furosemide permeability, since the results obtained with CACO-2 cells were in accordance with those derived with rat small intestine. However with CACO-2 monolayers, the reproducibility of the measurements was higher than with animal intestine due to the validated culture and experimental protocols, which ensured reproducible transporter expression and passive transcellular and paracellular membrane permeability in CACO-2 cells. Changes in paracellular permeability were easier to observe using CACO-2 cells than leakier cells or tissue (animal intestinal tissue); therefore, it was easier to elucidate all potential mechanisms underlying increased furosemide absorption.

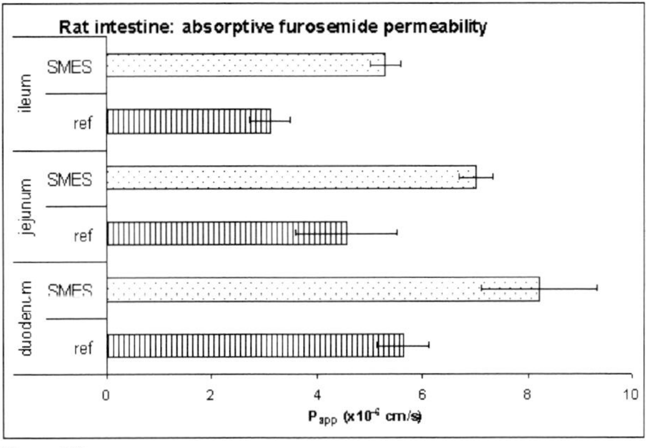

Figure 5A

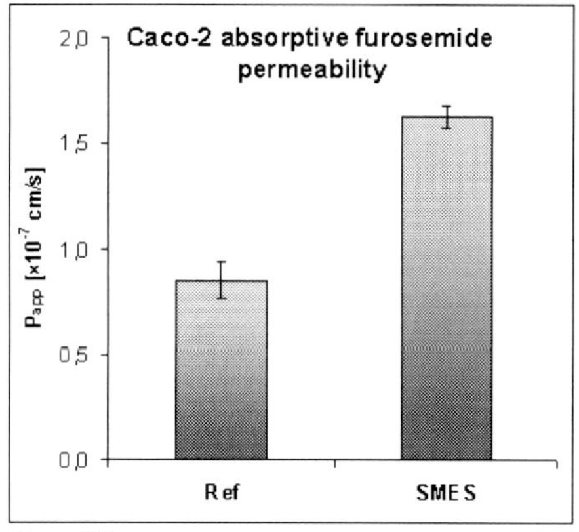

Figure 5B

Figure 5. The absorptive furosemide permeabilities through rat small intestine (duodenum, jejunum, and ileum) and through CACO-2 cell monolayers. In the presence of investigated SMES absorptive furosemide permeabilities significantly increased ($p < 0.05$) compared to the reference values.

CONCLUSION

Xenobiotics concomitantly ingested with prescribed drugs as food supplements (herbal supplements) or incorporated into dosage forms (lipids, surfactants) can impact pharmacokinetic behavior after per oral application by modifying a drug's physico-chemical characteristics (solubility, stability, release kinetics) or the function of biological barriers (metabolism, transport) and thus raise concerns regarding the efficacy and safety of the administered drugs. Trends in drug discovery and development tend to exploit simple, robust, reliable, low-cost and time-saving tools such as cellular models to perform mechanistic studies in pursuit of gaining full insight into the effects *in vivo*. These efforts have also received wide support from regulatory agencies, enforcing new guidelines (i.e. Guidance for Industry – Waiver of *in vivo* bioavailability and bioequivalence studies for immediate-release solid oral dosage forms based on biopharmaceutical classification system, Food-effect bioavailability and fed bioequivalence studies) on drug, excipient and formulation testing before the registration, which also include the aspect of using CACO-2 models.

In the presented contribution the applicability of the CACO-2 cell model as means to address the above stated claims has been presented. We have shown that pharmacokinetic interactions between garlic phytochemicals and different drugs could be correctly anticipated for non-BCS I compounds or drugs incorporated into modified release formulations. Interaction mechanisms were identified and their relevance to *in vivo* situation has been presented. Then the advantages of SMEDDS were investigated with CACO-2 cells. Molecular mechanisms of lipid constituents comprising SMEDDS that could improve the absorption of low soluble and/or low permeable drugs were elucidated. By showing our *in vitro* results, we have demonstrated the applicability and limitations of CACO-2 cells to test hypotheses pertaining to drug absorption.

REFERENCES

[1] Bailey, CA; Bryla, P; Malick, AW. *The use of the intestinal epithelial cell culture model, CACO-2, in pharmaceutical development.* Advanced Drug Delivery Reviews, 1996; 22: 85 – 103.

[2] Deferme, S; Annaert, P; Augustijns, P. *In vitro screening models to assess intestinal drug absorption and metabolism.* Drug absorption studies, Biotechnology: Pharmaceutical Aspects, 2008; Volume VII, 2: 182-215, DOI: 10.1007/978-0-387-74901-3_8.

[3] Lennernäs, H; Palm, K; Fagerholm, U; Artursson, P. *Comparison between active and passive drug transport in human intestinal epithelial (CACO-2) cells in vitro and human jejunum in vivo.* International Journal of Pharmaceutics, 1996; 127: 103 – 107.

[4] Artursson, P; Palm, K; Luthman, K. *CACO-2 monolayers in experimental and theoretical predictions of drug transport.* Advanced drug delivery reviews, 1996; 22: 67 – 84.

[5] Kommuru, TR; Gurley, B; Khan, MA; Reddy IK. *Self-emulsifying drug delivery systems (SEDDS) of Q10: formulation development and bioavailability assessment.* International Journal of Pharmaceutics, 2001; 212: 233 – 246.

[6] Gan, LSL; Thakker, DR. *Application of the CACO-2 model in the design and development of orally active drugs: elucidation of the biochemical and physical barriers posed by the intestinal epithelium.* Advanced Drug Delivery Reviews, 1997; 23: 77 – 98.

[7] Quaroni, A; Hochman, J. *Development of intestinal cell culture models for drug transport and metabolism studies.* Adnavced Drug Delivery Reviews, 1996; 22: 3 – 52.

[8] Pal, D; Mitra, AK. *MDR- and CYP3A4-mediated drug-herbal interactions.* Life Sciences, 2006; 78: 2131 – 2145.

[9] Volpe, DA. *Variability in CACO-2 and MDCK cell-based intestinal permeability assays.* Journal of Pharmaceutical Sciences, 2008; 97: 712 – 725.

[10] Hayeshi, R; Hilgendorf, C; Artursson, P; Augustijns, P; Brodin, B; Dehertogh, P; Fischer, K; Fossati, L; Hovenkamp, E; Korjamo, T; Masungi, C; Maubon, N; Mols, R; Müllertz, A; Mönkkönen, J; O'Driscoll, C; Oppers-Tiemissen, HM; Ragnarsson, EGE; Rooseboom, M; Ungell AL. *Comparrison of drug transporter gene expression and functionality in CACO-2 cells from 10 different laboratories.* European Journal of Pharmaceutical Sciences, 2008; 35: 383 – 396.

[11] Maubon, N; Le Vee, M; Fossati, L; Audry, M; Le Ferrec, E; Bolze, S; Fardel, O. *Analysis of drug transporter expression in human intestinal CACO-2 cells by real time PCR.* Fundamental & Clinical Pharmacology, 2007; 21: 659 – 663.

[12] Englund, G; Rorsman, F; Rönnblom, A; Karlbom, U; Lazorova, L; Gråsjö, J; Kindmark, A; Artursson, P. *Regional levels of drug transporters along human intestinal tract: Co-expression of ABC and SLC transporters and comparison with CACO-2 cells.* European Journal of Pharmaceutical Sciences, 2006; 29: 269 – 277.

[13] Seithel, A; Karlsson, J; Hilgendorf, C; Björquist, A; Ungell, AL. *Variability in mRNA expression of ABC and SLC transporters in human intestinal cells: Comparison between human segments and CACO-2 cells.* European Journal of Pharmaceutical Sciences, 2006; 28: 291 – 299.

[14] Sambuy, Y; De Angelis, I; Ranaldi, G; Scarino, ML; Stammati, A; Zucco, F. *The CACO-2 cell line as a model of intestinal barrier: influence of cell and culture-related factors on CACO-2 cell functional characteristics*. Cell Biology and Toxicology, 2005; 21: 1 – 26.
[15] Shah, P; Jogani, V; Bagchi, T; Misra, A. *Role of CACO-2 cell monolayers in prediction of intestinal drug absorption.* Biotechnol Prog, 2006; 22: 186 – 198.
[16] Hubatsch, I; Ragnarsson, EGE; Artursson, P. *Determination of drug permeability and prediction of drug absorption in CACO-2 monolayers.* Nature Protocols, 2007; 2: 2111 – 2119.
[17] Guidance for Industry – Waiver of in vivo bioavailability and bioequivalence studies for immediate-release solid oral dosage forms based on biopharmaceutical classification system *http://www.fda.gov/ cde /guidance/index.htm (accessed September 2010).*
[18] Avdeef, A; Tam, KY. *How well can the CACO-2/Madin-Darby Canine Kidney models predict effective human jejunal permeability?* Journal of Medicinal Chemistry, 2010; 53: 3566 – 3584.
[19] Yamashita, S; Furubayashi, T; Kataoka, M; Sakane, T; Sezaki, H; Tokuda, H. *Optimized conditions for prediction of intestinal drug permeability using CACO-2 cells.* European Journal of Pharmaceutical Sciences, 2000; 10: 195 – 204.
[20] Ingels, FM; Augustijns, PF. *Biological, Pharmaceutical, and Analytical considerations with the respect to the transport media used in the absorption screening system, CACO-2.* Journal of Pharmaceutical Sciences, 2003; 92: 1545 – 1558.
[21] Biganzoli, E; Cavenaghi, LA; Rossi, R; Brunati, MC; Nolli, ML. *Use of a CACO-2 cell culture model for the characterization of intestinal absorption of antibiotics.* Il Farmaco, 1999; 54: 594 – 599.
[22] Kawashiro, T; Yamashita, K; Zhao, XJ; Koyama, R; Tani, M; Chiba, K; Ishizaki, T. *A Study on the Metabolism of Etoposide and Possible Interactions with Antitumor or Supporting Agents by Human Liver Microsomes*. The Journal of Pharmacology and Experimental Therapeutics, 1998; 286: 1294 – 1300.
[23] Hande, KR; Krozely, MG; Greco, FA; Hainsworth, JD; Johnson, DH. *Bioavailability of low-dose oral etoposide*. Journal of Clinical Oncology, 1993; 11: 374 – 377.
[24] Shugarts, S; Benet, LZ. *The role of transporters in the pharmacokinetics of orally administered drugs*. Pharmaceutical Research, 2009; 26: 2039 – 2054.
[25] Proposal to waive in vivo bioequivalence requirements for the WHO model list of essential medicines immediate releaase, solid oral dosage forms. *WHO. http://www.who.int/medicines/services/exper tcommittees/ pharmprep/QAS04_109Rev1_Waive_invivo_bioequiv.pdf (accessed Dec. 2010).*
[26] Custodio, JM; Wu, CY; Benet, LZ. *Predicting drug disposition, absorption/elimination/ transporter interplay and the role of food on drug absorption*. Advanced Drug Delivery Reviews, 2008; 60: 717 – 733.
[27] Šturm, K; Žakelj, S; Vadnjal, L; Kristl, A. *Fluorescein transport properties across artificial lipid membranes, CACO-2 cell monolayers and rat jejunum*. European Journal of Pharmaceutics and Bio pharmaceutics, 2007; 66: 281 – 285.
[28] Balimane, PV; Chong, S. *Evaluation of permeability and P-glycoprotein interactions: industry outlook, chapter 5*: Biopharmaceutics application in drug development, editors Krishna R, Yu L, Springer 2008.

[29] Hidalgo, IJ; Li, J. *Carrier-mediated transport and efflux mechanisms in CACO-2 cells.* Advanced Drug Delivery Reviews, 1996; 22: 53 – 66.

[30] Dietrich, CG; Geier, A; Oude Elferink, RPJ. *ABC of oral bioavailability: transporters as gatekeepers in the gut.* Gut, 2003; 52: 1788 – 1795.

[31] Pal, D; Mitra, AK. *MDR- and CYP3A4-mediated drug-herbal interactions.* Life Sciences, 2006; 78: 2131 – 2145.

[32] Ioannides, C. *Drug-phytochemical interactions.* Inflammopharmacology, 2003; 11: 7 – 42.

[33] Amagase, H. *Clarifying the real bioactive constituents of garlic.* Journal of Nutrition, 2006; 136: 716S – 725S.

[34] Barnes, J; Anderson, LA; Phillipson, J. *Herbal medicines.* 3rd ed. London Chicago, Pharmaceutical Press 2007.

[35] Venkataramanan, R; Komoroski, B; Strom, S. *In vitro and in vivo assessment of herb drug interactions.* Life Sciences, 2006; 78: 2105 – 2115.

[36] Klotz, U. *Pharmacokinetics and drug metabolism in elderly.* Drug Metabolism Reviews, 2009; 41: 67 – 76.

[37] Drozdowski, L; Thomson, ABR. *Aging and the intestine.* World Journal of Gastroenterology, 2006; 12: 7578 – 7584

[38] Ahmed, T; Haboubi, N. *Assessment and management of nutrition in older people and its importance to health.* Clinical Interventions in Aging, 2010; 5: 207 – 216.

[39] Agarwal, KC. *Therapeutic actions of garlic constituents.* Medicinal Research Reviews, 1996; 16: 111 – 124.

[40] Ejaz, S; Woong, LC; Ejaz, A. *Extract of garlic (Allium sativum) in cancer chemoprevention.* Experimental Oncology, 2003; 25; 93 – 97.

[41] Zhou, S; Gao, Y; Jiang, W; Huang, M; Xu, A; Paxton, JW. *Interactions of herbs with Cytochrome P450.* Drug Metabolism Reviews, 2003; 35: 35 – 98.

[42] Borrelli, F; Capasso, R; Izzo, AA. *Garlic (Allium sativum L.): adverse effects and drug interactions in humans.* Molecular Nutrition & Food Research, 2007; 51: 1386 – 1397.

[43] Abebe, W. *Herbal medication: potential for adverse interactions with analgesic drugs.* Journal of Clinical Pharmacy and Therapeutics, 2002; 27: 391 – 401.

[44] Banerjee, SK; Maulik, SK. *Effect of garlic on cardiovascular disorders: a review.* Nutrition Journal, 2002; 1: 4 – 18.

[45] Hu, Z; Yang, X; Ho, PCL; Chan, SY; Heng, PWS; Chan, E; Duan, W; Koh, HL; Zhou, S. *Herb-drug interactions, a literature review.* Drugs, 2005; 65: 1239 – 1282.

[46] Izzo, AA. *Herb-drug interactions: an overview of the clinical evidence.* Fundamental & Clinical Pharmacology, 2004; 19: 1 – 16.

[47] Williamson, EM. *Drug interactions between herbal and prescribed medicines.* Drug Safety, 2003; 26: 1075 – 1092.

[48] Mills, E; Montori, V; Perri, D; Phillips, E; Koren, G. *Natural health-HIV drug interactions: a systematic overview.* International Journal of STD & AIDS, 2005; 16: 181 – 186.

[49] Fugh-Bergman, A; Ernst, E. *Herb-drug interactions: review and assessment of report reliability.* British Journal of Clinical Pharmacology, 2001; 52: 587 – 595.

[50] Coecke, S; Rogiers, V; Bayliss, M; Castell, J; Doehmer, J; Fabre, G; Fry, J; Kern, A; Westmoreland, J. *The use of long-term hepatocyte cultures for detecting induction of*

drug metabolizing enzymes: the current status. Alternatives to Laboratory Animals, 1999: 27: 579 – 638.

[51] Brandon, EFA; Raap, CD; Meijerman, I; Beijnen, HS; Schellens, JHM. *An update on in vitro methods in human hepatic drug biotransformation research: pros and cons.* Toxicology and Applied Pharmacology, 2003; 189: 233 – 246.

[52] Piscitelli, SC; Burnstein, AH; Welden, N; Gallicano, KD; Falloon, J. *The effect of garlic supplements on the pharmacokinetics of saquinavir.* Clinical Infectious Diseases, 2002; 34: 234 – 238.

[53] Berginc, K; Žakelj, S; Kristl, A. *In vitro interactions between aged garlic extract and drugs used for the treatment of cardiovascular and diabetic patients.* European Journal of Nutrition, 2010; 49(6): 373 – 384.

[54] Berginc, K; Trdan, T; Trontelj, J; Kristl, A. *HIV protease inhibitors: garlic supplements and first-pass intestinal metabolism impact on the therapeutic efficacy.* Biopharmaceutics & drug dispotition, 2010; 31: 495 - 505.

[55] Tang, C; Prueksaritanont, T. *Use of in vivo animal models to assess pharmacokinetic drug-drug interactions.* Pharmaceutical Research, 2010; 27: 1772 – 1787.

[56] Taipalensuu, J; Törnblom, H; Lindberg, G; Einarsson, C; Sjöqvist, F; Melhus, H; Garberg, P; Sjöström, B; Lundgren, B; Artursson, P. *Correlation of gene expression of ten drug efflux proteins of the ATP-binding cassette transporter family in normal human jejunum and in human intestinal epithelial CACO-2 cell monolayers.* The Journal of Pharmacology and Experimental Therapeutics, 2001; 299: 164 – 170.

[57] Haimeur, A; Conseil, G; Deeley, RG; Cole, SP. *The MRP-related and BCRP/ABCG2 multidrug resistance proteins: biology, substrate specificity and regulation.* Current Drug Metabolism, 2004; 5: 21 – 53.

[58] Masereel, B; Pochet, L; Laeckmann, D. *An overview of inhibitors of Na/H exchanger.* European Journal of Medicinal Chemistry, 2003; 38: 547 – 554.

[59] Anderle, P; Huang, Y; Sadee, W. *Intestinal membrane transport of drugs and nutrients: genomics of membrane transporters using expression microarrays.* European Journal of Pharmaceutical Sciences, 2004; 21: 17 – 24.

[60] Martin, C; Berridge, G; Higgins, CF; Mistry, P; Charlton, P; Callagan, R. *Communication between multiple drug binding sites on P-glycoprotein.* Molecular Pharmacology, 2000; 58: 624 – 632.

[61] Gerk, PM; Vore, M. *Regulation of expression of the multipledrug resistancee-associated protein 2 (MRP-2) and its role in drug disposition.* Journal of Pharmacology & Experimental Therapeutics, 2002; 302: 407 – 415.

[62] Steffansen, B; Nielsen, CU; Brodin, B; Eriksson, AH; Andersen, R; Frokjaer, S. *Intestinal solute carriers: an overview of trends and strategies for improving oral absorption.* European Journal of Pharmaceutical Sciences, 2004; 21: 3 – 16.

[63] Berginc, K; Žakelj, S; Uršič, D; Kristl, A. *Aged garlic extract stimulates P-glycoprotein and Multidrug resistance associated protein 2 mediated effluxes.* Biological & Pharmaceutical Bulletin, 2009; 32: 694 – 699.

[64] Shugarts, S; Benet, LZ. *The role of transporters in the pharmacokinetics of orally administered drugs.* Pharmaceutical Research 2009; 26: 2039 – 2054.

[65] Benet, LZ. *Using the BDDCS to predict when metabolism and drug transport influence the disposition of a drug*. AAPS workshop on drug transporters in ADME: from the bench to the bedside, March 8, 2005.
[66] Wang, YH; Chao, PDL; Hsiu, SL; Wen, KC; Hou. YC. *Lethal quercetin-digoxin interactions in pigs*. Life Sciences, 2004; 74: 1191 – 1197.
[67] Langman, P; Zilly, M; Winzer, R; Klinker, H. *Therapeutic drug monitoring: a tool to individualize highly active antiretroviral therapy in HIV infected patients.* Current Pharmaceutical Analysis, 2006; 2: 205 – 217.
[68] Petit, F; Fromenty, B; Owen, A; Estaquier, J. *Mitochondria are sensors for HIV drugs.* Trends in Pharmacological Sciences, 2005; 26: 258 – 264.
[69] Zvonar, A; Berginc, K; Kristl, A; Gašperlin, M. *Microencapsulation of self-microemulsifying system: improving solubility and permeability of furosemide.* International Journal of Pharmaceutics, 2010; 388: 151 – 158.
[70] Xianyi, S; Guijun, Y; Yunjuan, W; Juncan, L; Xiaoling, F. *Effect of self-emulsifying drug delivery systems containing Labrasol on tight junctions in CACO-2 cells.* European Journal of Pharmaceutical Sciences; 2005; 24: 477 – 486
[71] Daham, A; Hoffman, A. *The effects of different lipid based formulations on the oral absorption of lipophilic drugs: the ability of in vitro lipolysis and consecutive ex vivo intestinal permeability data to predict in vivo bioavailability in rats.* European Journal Pharmaceutics and Biopharmaceutics, 2007; 67: 96 – 105.
[72] Perng, CH; Kearney, AS; Patel, K; Palepu, NR; Zuber, G. *Investigation of formulation approaches to improve dissolution of SB-210661, a poorly soluble 5-lipooxigenase inhibitor.* International Journal of Pharmaceutics, 1998; 176: 31 - 38.
[73] Homar, M; Dreu, R; Kerč, J; Gašperlin, M. *Preparation and evaluation of celecoxib-loaded microcapsules with self-microemulsifying core.* Journal of Microencapsulation, 2009; 26: 479 - 484.
[74] Zvonar, A; Rozman, B; Bešter-Rogač, M; Gašperlin, M. *The influence of microstructure on celecoxib release from a pharmaceutically applicable system: Mygliol 812/Labrasol/Plurol oleiquer/water mixtures.* Acta Chimica Slovenica, 2009; 56: 131 - 138.
[75] Dolenc, A; Govedarica, G; Dreu, R; Kocbek, P; Srčič, S; Kristl, J. *Nanosized particles of orlistat with enhanced in vitro dissolution rate and lipase inhibition.* International Journal of Pharmacetics, 2010; 396: 149 – 155.
[76] Porter, CJH; Trevaskis, NL; Charman, WN. *Lipids and lipid-based formulations: optimizing the oral delivery of lipophilic drugs.* Nature Reviews Drug Discovery, 2007; 6: 231 - 248.
[77] Pouton, CW. *Lipid formulations for oral administration of drugs: non-emulsifying, self emulsifying, and `self microemulsifying` drug delivery systems.* European Journal of Pharmaceutical Sciences, 2000; 11 (Suppl. 2): S93 - S98.
[78] Pouton, CW. *Formulation of poorly water-soluble drugs for oral administraniton: physicochemical and physiological issues and the lipid formulation classification system.* European Journal of Pharmaceutical Sciences, 2006; 29: 278 - 287.
[79] Shah, NH; Carvajl, MT; Patel, PI; Infeld, MH; Malick, AW. *Self-emulsifying drug delivery systems (SEDDS) with polyglycolized glycerides for improving in vitro dissolution and oral absorption of lipophilic drugs.* International Journal of Pharmaceutics, 1994; 106: 15 - 23.

[80] Constantinides, PP. *Lipid microemulsions for improving drug dissolution and oral absorption: Physical and biopharmaceutical aspects.* Pharmaceutical Research, 1995; 12(11): 1561 - 1572.

[81] Neslihan Gursoy, R; Benita, S. *Self-emulsifying drug delivery systems (SEDDS) for improved oral delivery of lipophilic drugs.* Biomedicine & Pharmacotherapy, 2004; 58: 173 - 182.

[82] Hauss, DJ. *Oral lipid-based formulations.* Advanced Drug Delivery Reviews, 2007; 59: 667 - 676.

[83] Pouton, CW; Porter, CJH. *Formulation of lipid-based delivery systems for oral administration: materials, methods and strategies.* Advanced Drug Delivery Reviews, 2008; 60: 625 - 637.

[84] O`Driscoll, CM; Griffin, BT. *Biopharmaceutical challanges associated with drugs with low aqueous sulubility-the potential impact of lipid-based formulations.* Advanced Drug Delivery Reviews, 2008; 60: 617 - 624.

[85] Mei-Ling, C. *Lipid excipients and delivery systems for pharmaceutical development: a regulatory perspective.* Advanced Drug Delivery Reviews, 2008; 60: 768 - 777.

[86] Seebaluck, F; Ashhord, MB; O`Driscoll, CM. *The effects of pluronics block copolymers and cremophor EL on intestinal lipoprotein processing and the potential link with P-glycoprotein in CACO-2 cells.* Pharmaceutical Research, 2003; 20: 1085 - 1092.

[87] Gajjar, R; Lo, CM; Tso, P. *Physiological processes governing gastrointestinal absorption of lipids and lipophilic xenobiotics, in: Hauss DJ(Ed.), Oral lipid-based formulations: enhancing the bio availability of poorly water soluble drugs,* Informa Healthcare, Inc., New York, 2007, pp. 207 – 239.

[88] Trevaskis, NL; Charman, WN; Porter, CJH. *Lipid-based delivery systems and intestinal lymphatic drug transport: a mechanistic update.* Advanced Drug Delivery Reviews, 2008; 60: 702 - 716.

[89] Dahan, A; Hoffman, A. *Rationalizing the selection of oral lipid based drug delivery systems by an in vitro dynamic lipolysis model for improved oral bioavailability of poorly water soluble drugs.* Journal of Controlled Release, 2008; 129: 1 - 10.

[90] Buyukozturk, F; Benneneyan, JC; Carrier, RL. *Impact of emulsion-based drug delivery systems on intestinal permeability and drug release kinetics.* Journal of Controlled Release, 2010; 142: 22 – 30.

[91] O'Driscoll, CM. *Lipid-based formulations for intestinal lymphatic delivery. Mini review.* European Journal of Pharmaceutical Sciences, 2002; 15: 405 – 415.

[92] Grove, M. *Lipid Self-microemulsifying drug delivery systems, in: Hauss DJ (Ed.), Oral lipid-based formulations: enhancing the bioavailability of poorly water soluble drugs,* Informa Healthcare, Inc., New York, 2007, pp. 107 - 127.

[93] Fatouros, DM; Karpf, DM; Nielsen, FS; Mullertz, A. *Clinical studies with oral lipid based furmulations of poorly soluble compounds.* Therapeutics and Clinical Risk Management, 2007; 3(4): 591 - 604.

[94] Chakraborty, S; Shukla, D; Mishra, B; Singh, S. Lipid – *An emerging platform for oral delivery of drugs with poor bioavailability.* European Journal of Pharmaceutics and Biopharmaceutics, 2009; 73: 1 - 15.

[95] O'Driscoll, CM. *Intestinal lymphatic targeting of drugs.* STP Pharma Sciences, 2003; 13: 17 – 25.

[96] Sha, X; Yan, G; Wu, Y; Li, J; Fang, X. *Effect of self-microemulsifying drug delivery systems containing Labrasol on tight junctions in CACO-2.* European Journal of Pharmaceutical Sciences, 2005; 24: 477 – 486.

[97] Cano-Cebrián, MJ; Zornoza, T; Granero, L; Polache, A. *Intestinal absorption enhancement via the paracellular route by fatty acids, chitosan and others: a target for drug delivery.* Current Drug Delivery, 2005; 2: 9 – 22.

[98] Constantinides, PP; Kishor, M; Wasan, KM. *Lipid formulation strategis for enhancing intestinal transport and absorption of P-glycoprotein (P-gp) substrate drugs: in vitro/in vivo case studies. Review.* Journal of Pharmaceutical Sciences, 2007; 96: 235 – 248.

[99] Goole, J; Lindley, DJ; Roth, W; Carl, SM; Amighi, K; Kauffmann, JM; Knipp, GT. *The effects of excipients on transporter mediated absorption. Review.* Internation Journal of Pharmaceutics 2010; 393: 17 – 31.

In: CACO-2 Cells and their Uses
Editor: Megan A. Schulz, pp. 31-48
ISBN 978-61209-917-0

Chapter 2

A Critical Perspective on Cell Lines Studies in Nutrition: The Case of Intestinal Absorption

C. Frontela-Saseta[1], P. Peso-Echarri[2], C. González-Bermúdez[2], R. López-Nicolás[2], C. Martínez-Graciá[2] and G. Ros-Berruezo[2]

[1]National Research Institute for Food and Nutrition, Inran, Rome, Italy
[2]Department of Food Science and Nutrition
Faculty of Veterinary Sciences, 30071 Murcia, Spain

Abstract

In nutrition research, *in vitro* cellular models have been considered as an expedient approach to the study of biological mechanisms underlying the effect of molecules of nutritional interest on different physio-pathological contexts and as the method of choice to obtain observations that can be extrapolate to *in vivo* studies. However, some critical points still need to be elucidated to land to definitive conclusions on the applicability of specific cell lines for the understanding of mechanisms associated with physiological processes. Although significant progress has been made throughout the time, it is generally recommended to take experimental results on cultured cells wisely and with caution. In this review we outline some aspects related with the useful of cell lines in intestinal absorption studies, including the identification of the experiment's objective, molecular mechanisms and biological aspects involved at cellular level and the interpretation of results that can be elucidated by *in vitro* assays, including approaches recently developed. We will mainly focus on specific cell lines used as a model for intestinal absorption, discussing the "pros and cons" and the critical aspects to be considered while planning a study. The value and possible limitations of the utilization of some specific cell lines will be presented and discussed, commenting on some aspects of the clinical predictive value of these *in vitro* models.

INTRODUCTION

It has been widely reported that dietary components are able to affect different aspects of the physiological and pathological cellular response significantly, corroborating the evidence of a strong link between nutrition and health. Acquiring a deeper understanding of the molecular mechanisms involved in the nutrition process often involves *in vivo* assays. Animal and human models are invaluable tools for these nutritional investigations, as they provide epidemiological data on the consumption of certain foods and the risk of cancer, the appearance of certain inflammatory processes and other important diseases [1].

The European Centre for the Validation of Alternative Methods [2] promotes the scientific and regulatory acceptance of alternative methods that reduce, refine or replace the use of laboratory animals with the aim of validating the possible incorporation of non-animal tests into regulatory procedures. Frequently, *in vitro* systems are designed to isolate the barrier of interest to permit relatively rapid evaluation of candidate compounds. However, knowledge of the routes and mechanisms of transport of molecules across the intestinal epithelium is crucial when evaluating absorption [3], which depends on the physicochemical properties of the absorbed compound. Moreover, intestinal absorption studies must consider physiological factors such as gastrointestinal motility, pH, blood flow, nutrition, pathological state and mucus dissolution among other factors. However, considering that none of the cell models established, up to now, can fully resemble intestinal absorption *in vivo*, extrapolation of the data obtained by *in vitro* cell systems to the *in vivo* situation is difficult [4]. During the intestinal absorption process, compounds passing through the intestinal epithelium (involving both passive diffusion and active transport into the blood or lymph) may be metabolized, appearing at the basal surface of intestinal epithelial cells in a changed form [4]. Thus, it is crucial to select an appropriate model depending on the test compounds.

RESEARCH WITH CELL LINES

Initially, cell cultures (mainly obtained from tumour tissues) were considered an expedient model for the study of the cellular and molecular aspects of cancer [5]. However, more recently, tumour cells have been demonstrated to be an adequate tool to study "normal" physiological and non-pathological processes at the cellular level, being pivotal in the characterization of the activity of small molecules and in the screening of metabolic compounds [6,7]. Evidence in the literature suggests that cancer cell lines are also a valuable tool to characterize the effects of some components of foods on some metabolic processes [8].

The use of cell lines to study absorption processes at the intestinal level requires not only the use of the adequate cell type, but also a deep knowledge of the physiological conditions involved in order to replicate *in vivo* intestinal conditions. In this regard, it has been reported that different pH values during experiments with cell lines have a significant effect on the *in vitro* permeability of dietary compounds [9,10]. During *in vivo* intestinal digestion, the luminal pH value plays a critical role in nutrient absorption and bioavailability. It has been reported that, under the condition of a luminal pH value of 6.0, better predictions of *in vivo* drug absorption in humans are obtained [11]. Due to this fact, during *in vitro* cell assays, the pH value of the apical and basolateral medium must be carefully tested before and during experiments. Tsuji et al. [12] reported the pH dependence of the transcellular transport of certain organic acids across the human colon carcinoma cell line CACO-2 (the most extensively characterized and frequently used of the available cell lines [13]). It must also be noted that the functional activity of the P-glycoprotein (P-gp) transporter (strongly expressed in CACO-2 cells), which plays a key role in intestinal absorption, is affected by pH-dependent passive permeability [14].

During studies of intestinal absorption by cell lines, it is also important to consider the effects of bile acids, surfactant and some agents used in the *in vitro* gastrointestinal digestion process in order to obtain a satisfactory *in vitro–in vivo* correlation [11]. Moreover, to simulate *in vivo* conditions in the digestive tract, peristaltic movements and the different conditions on each part of the gastrointestinal tract should be reproduced by an *in vitro* dynamic gastrointestinal model to resemble gastrointestinal conditions. Therefore, as reported by Verwey et al. [15], this system coupled to a human colon carcinoma cell line could be an important tool to study the transport and kinetics of dietary compounds through the intestinal wall. The human intestinal lining is covered by a layer of mucus. Although it has been reported that the existing *in vitro* digestion/CACO-2 cell culture model correlates well with human bioavailability studies, in the case of iron, the mucus appears to play a role in intestinal absorption [16]. To simulate this physiological condition, the use of a second type of cell line (one that produces a mucus layer over the CACO-2 intestinal cells grown in culture), has been proposed. This closely imitates the conditions in the body to see if the addition of the mucus layer changes intestinal absorption in the cells, giving more accurate iron bioavailability predictions [17].

The study of nutrient fate and metabolism has benefited from the development of *in vitro* studies by cell lines, providing information about the turnover of some nutrients as well as about energy metabolism in different cells and tissues by means of appropriate markers and tracers (i.e., enzymatic changes) [18]. The kinetics of different metabolic processes (glycolysis, protein synthesis, lipolysis, etc.) can also be evaluated by *in vitro* assays that provide valuable information about nutrient turnover and functions.

In particular, entering into the matter of this review, cultured cells have been utilized to obtain a better understanding of nutritional processes at the intestinal level. They are an empirical but valuable system for the identification and characterization of the interactions between different molecules. They are also significant due to their role as mediators in intestinal absorption and their involvement in specific physiological processes at a cellular level in specific cell lines [19]. These observations were key to elucidating the kinetics of drug intestinal transport and metabolism, to understanding the effect of dietary compounds on the intestinal wall, to studying some aspects of nutrient transport [20] and to assessing which dietary components affect the intestinal absorption process of essential molecules [21]. Moreover, cell lines have allowed the study of the effect of isolated food components on regulatory genes and on cell cycle progression. Knowledge of the effects of specific molecules has also improved due to enhanced control of environmental conditions during experiments. Meanwhile in *in vivo* models, the variables involved in the process of absorption cannot be separated, making it impossible to identify individual rate-limiting factors [2]. The objectives of *in vitro* studies with cell lines are likely to be met only if studies are designed in a way that is consistent with meeting with those objectives, articulating clearly what investigators hope to get out of cell studies.

Cell Lines in Intestinal Absorption Studies

Some intestinal cell lines are intensively used in nutrition research; however, the use of some of them as *in vitro* models for intestinal absorption is limited because of their different properties. It should be noted that there are studies of absorption mechanisms reporting that a model containing only absorptive cells without interferences of mucus, lamina propria or muscularis mucosa is a well-performed *in vitro* model to study intestinal absorption mechanisms [2]. However, data obtained from these cell models must be taken considering that an adequate choice of cell line is crucial. In this regard, in intestinal absorption studies, attention has turned to the use of human carcinoma cell lines that display properties characteristic of different intestinal cells. Human adenocarcinoma cell lines have been proposed as the better choice to reproduce the human intestinal absorption process. The main intestinal cell lines of human origin used in scientific research are described below:

CACO-2 is the most widely used cell line as a model of human intestinal absorption. This cell line is derived from a human colorectal adenocarcinoma and differentiates spontaneously to cells that resemble mature human small intestinal enterocytes. As a monolayer, CACO-2 cells differentiate to form tight junctions between cells. This could also make it useful as a model of paracellular movement of compounds across the monolayer. The differentiation is completed within 14-21 days after seeding [22] (Figure 1). This cell line has been demonstrated to express important active transport proteins, efflux and phase II conjugation enzymes and nuclear receptor keys in the intestinal absorption mechanisms to model a variety of transcellular pathways as well as metabolic transformation of test substances. It also exhibits high levels of alkaline phosphatase, sucrose-isomaltase and aminopeptidase activity characteristic to enterocyte brush border microvilli. Because of these properties, the CACO-2 cell line has great potential for intestinal absorption studies of some dietary compounds, mainly for studying mechanism transport [23]. It must be considered that *in vitro* techniques require determination of apparent permeability (Papp) when studying transport across cell monolayer, considering it as a single barrier [24]. A large number of (expressed as apparent permeability coefficient – Papp value) values for different molecules are being produced and are/will be available to build databases with a large span of values from high to low bioavailabilities. In nutrition research, CACO-2 cells have been and are a very usable tool to study intestinal absorption of different substances (i.e., absorption and bioavailability of carotenoids) [25], effects of pre- and probiotics on Fe availability [26], intestinal absorption of Fe in vegetarians [27], absorption of furanones of fruit and vegetables [28] and absorption of glucose [29]. The study of the apparent permeability of these molecules can be assessed *in vitro* by a bicameral system of cells cultured on permeable supports where the flux of transport can be easily measured [30].

However, the use of the CACO-2 cell line in absorption experiments may have some limitations and disadvantages due to its low paracellular permeability compared to that of the human small intestine [13]. This is probably due to the colonic origin of the cells or the fact that the tight junctions display a tightness commonly found in the colonic epithelium. Many active transport proteins of the CACO-2 cells are expressed in different quantities in the small intestine. Moreover, the lack of the crypt-villus axis in the CACO-2 cell line, which is important for fluid and ion transport *in vivo*, as well as the lack of mucus production could also be considered significant disadvantages when this cell line is used in intestinal absorption assays [31]. The permeability glycoprotein (P-gp), extensively expressed in the intestinal epithelium, exhibits an unexpectedly low efflux ratio in CACO-2 cells, limiting the usefulness of this cell line for intestinal absorption studies. CYP3A4 is the most abundant cytochrome P450 (a group of monooxygenases of particular importance when studying drug metabolism) present in the epithelial cells (enterocytes) that line the lumen of the small bowel [32] with an important barrier function against xenobiotics. However, this protein is very weakly expressed in CACO-2 cells. Based on this fact, some alternatives have been proposed. For example, treatment of CACO-2 cells with 1α,25-dihydroxyvitamin D3 has been proposed to increase the endogenous CYP3A4 expression levels [33]. Crespi et al. (2000) [34] introduced CYP3A4 in the CACO-2 cell line by DNA manipulations to study drug transport in the cell monolayer system. It has also been demonstrated that after exposure to methotrexate, a subclone of CACO-2 cells (called TC-7 with a stable expression of CYP3A4) is obtained, offering marked advantages over the parental CACO-2 cell line [35]. In addition, TC-7 has shown a good correlation coefficient of permeability for many drugs compared with CACO-2 cells [36], making TC-7 a reliable alternative to CACO-2. In an attempt to approach *in vivo* intestinal conditions, co-cultures of CACO-2 cells and mucus-producing goblet cells have been proposed as an improved model of intestinal absorption [4]. These weaknesses determine the need to establish an adequate relationship between permeability data obtained in CACO-2 cells and predictions of absorption in humans. Despite the existence of studies [37,38, 39,40] informing that the *in vitro* digestion/CACO-2 cell culture model for intestinal absorption of different compounds correlates well, qualitatively, with human data, up to date, no satisfactory results have been observed. Moreover, there have been important discrepancies between laboratories, determining that CACO-2 cells alone are not useful in predicting human intestinal absorption quantitatively (ECVAM Workshop, 2001).

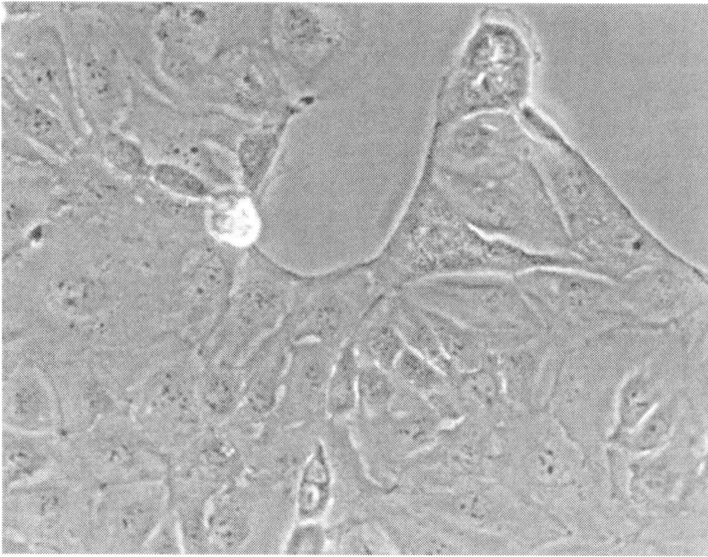

Figure 1. (Continued).

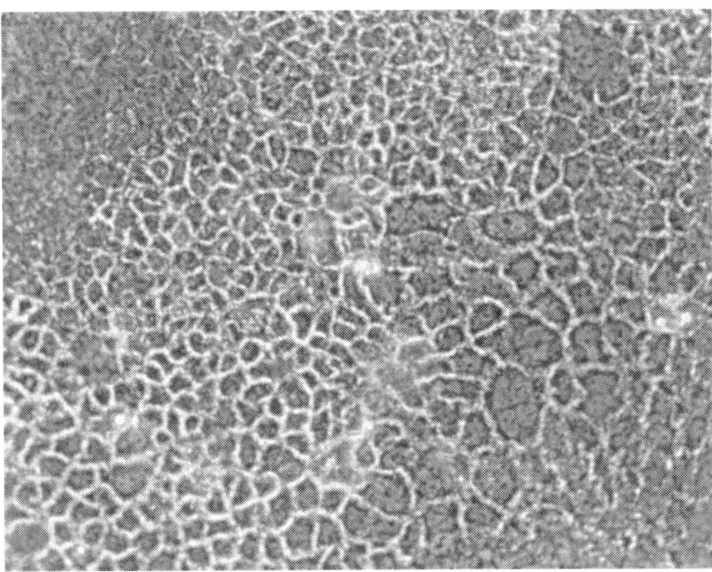

Figure 1. CACO-2 cells after 4 days postseeding (A) and after 19 days postseeding (B) (differentiated cells forming a monolayer).

Since the use of CACO-2 cells in absorption studies has grown exponentially in recent years, it is particularly important to highlight these methodological aspects in order to promote the standardization and optimization of this intestinal model as well as a better knowledge of cell lines that could be adequate alternatives. It must be noted that CACO-2 cells have been shown to be a heterogeneous population and highly dependent on culture conditions; moreover, the CACO-2 cell line does not always express appropriate amounts of transporters or enzymes, which may determine bias in the determination of nutrient transport via the carrier-mediated process or their metabolism via a particular pathway. More recently, CACO-2 cells have been used to determine the efflux mechanisms of phase II conjugates of drugs and natural products [41]. Based on the information reported, additional genetic manipulation of the CACO-2 cells is needed to advance the utility of this model in the study of intestinal absorption mechanisms further. Based in this fact, the improvement of this cell line as a model of intestinal permeability and transport by genetic modification has led to the development of different clones that are reviewed in this chapter.

The TC7 cell line, a clone isolated from a late passage of the parental CACO-2 cell line, has proven to be a good option for studying intestinal passive diffusion, because TC7 cells have been observed to correlate well with the parental cell line [42]. However, TC7 cells are unsuitable to predict intestinal absorption (i) for highly lipophilic compounds, (ii) for poorly absorbed compounds or (iii) when transporter-mediated routes and/or first pass metabolism are involved [42]. Moreover, the beneficial effect of phenolic compounds on health cannot be understood without studying its bioavailability or the fraction absorbed by the intestinal cells. Soler et al. (2010) [43] have used the human small-intestinal epithelial CACO-2/TC7 cell line to study the intestinal transport of olive oil phenols, informing that this model of the human intestinal epithelium offers limited metabolism of phenolics. However, these authors also informed that phenols were absorbed inside the enterocyte and conjugated to different degrees in this co-culture model, thus making it a good tool for understanding the absorption of phenols. Zucco et al. (2005) [44] reported that the TC7 clone consists of a more homogeneous population, giving rise to less variable results and exhibiting more developed intercellular junctions than other clones of CACO-2 cells. However, due to the lack of knowledge regarding the characterization of the TC7 cellular model with respect to metabolic and transporter competence and the lack of knowledge regarding whether this cell clone is a good model for studying intestinal absorption, more studies are needed.

A new cell line, HCA-7, has been established from a primary human colonic adenocarcinoma. The HCA-7 cells have an epithelial morphology by phase-contrast microscopy. HCA-7 cells retain some of the morphological and functional polarity exhibited by normal colonic epithelium. Cells form a polarized epithelial sheet when grown on tissue culture plastic in standard culture medium. The cells are shown to be structurally polarized with sparse microvilli present at the apical surface but absent from the basolateral surface. Cells are connected by apical tight junctions and frequent desmosomes [45]. In an attempt of approach to the absorption mechanisms in the gastrointestinal tract, Buck et al. (2006) [46] developed an *in vivo*-like *in vitro* intestinal epithelial model by co-culturing human intestinal myofibroblasts and intestinal epithelial cell lines (CACO-2 and HCA-7). According to these authors, this model represents a novel way of recreating the intestinal environment *in vitro*. The mucus layer covering the intestinal epithelium is thought to play a role in the absorption of some nutrients such as iron. Mucus is secreted by goblet cells in the intestinal epithelium and is composed by heavily glycosylated proteins, which are responsible for the role of mucin in iron uptake.

The HT29 cell line, derived from human colon carcinoma, has been proposed as an alternative to CACO-2 cells or as a way to improve it in intestinal absorption models, since this human cell line from colon carcinoma, cultured in the presence of galactose, expresses moderate enterocytic differentiation [47] (Figure 2).

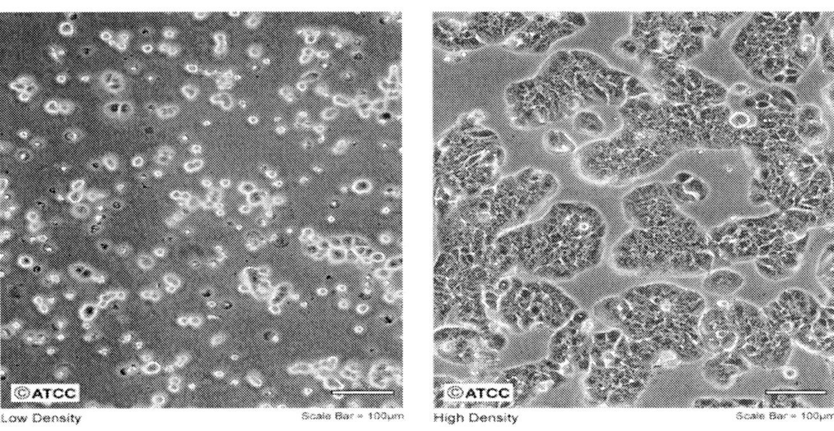

Figure 2. HT-29 cells seeded at low (left) and high (right) density (obtained from the American Type Culture Collection (ATCC)).

Several HT-29 clones differentiate into enterocytic cells or goblet cells, which secrete mucus. Between them, two clones that secrete mucus are HT-29-H and HT-29-MTX. The latter is obtained by adapting the cells to a medium containing methotrexate. HT29-MTX expresses the morphological and mucin-producing characteristics of goblet cells (Laparra et al., 2009). In an attempt to incorporate into CACO-2 cell monolayers characteristics more representative of the intestinal mucosa, a co-culture system consisting of absorptive and mucus-secreting cell types has been established by CACO-2/HT29-MTX, correlating well with absorption data in humans [4]. Some dietary compounds have also been studied using HT29 cells (i.e., the effect of fermented milks on colon cancer) [48]. HT29 cells have also been demonstrated to be useful for studying the effect of antioxidant compounds (i.e., anthocyanidins) from foods preventing tumoural processes and for clarifying the mechanisms in which some dietary components are involved in mediating inflammatory intestinal processes [49]. A recent study [50] about mineral carriers (caseinphosphopeptides) at the intestinal level has demonstrated that both cell lines (CACO-2 and HT29), when differentiated, are good tools for studying calcium uptake. However, similar to the CACO-2 cell line, HT29 cell monolayers are not a perfect model of all transport routes in the normal intestinal enterocyte because of their low paracellular permeability compared to that of the human small intestine [13]. This likely indicates that the co-culture of CACO-2 and HT29-MTX could be more usable for intestinal absorption studies that use each intestinal cell line separately.

Other tumoral intestinal cell lines of human origin such as HCT-116 and SW620 (widely used in the study of the anticarcinogenic effects of some dietary compounds) that are not acceptable for evaluation of absorption properties have been reported. These have not been shown to express proteins important for regulating intestinal absorption [51].

NCM 460 is a normal human colon mucosal epithelial cell line that has proved useful in vitamin intestinal absorption research on transport, gene regulation and protein expression [52]. Some of the cells are positive for mucin synthesis. This cell line has been chosen mainly for studying vitamin absorption, because it possesses characteristics similar to that of the native colonocytes, including similar uptake mechanisms with the existence of specialized mechanisms of uptake for riboflavin, biotin and thiamine [52,53]. Similar to CACO-2 cells, the human colonic epithelial cell line, LS174T, provides an excellent *in vitro* model to elucidate the cell-type specific mechanisms responsible for mucin expression that play a key role in intestinal absorption mechanisms [54].

Non-Human Cell Lines in Intestinal Absorption Studies

Many non-human intestinal epithelial cell lines have been used as *in vitro* models of intestinal absorption. According to certain studies, the rat origin cell line IEC-18 could be a good option to study transport routes in the small intestine, since these cells are derived from ileum [55]. Natural compounds that are transported paracellularly would probably be inaccessible in the CACO-2 cell model. In this regard, the IEC-18 cell line seems to be more adequate to study paracellular transport [56]. Many nutrient absorption studies, using this cell line, have been extrapolated to human intestinal absorption studies. At the intestinal level, this cell line expresses iron transport proteins involved in the uptake and efflux of iron [57]. It has been also used to study the absorption of fatty acids, showing specific transporters that make it useful to study vitamin absorption at the intestinal level [58]. Pérès et al. (1999) [59] studied the mechanisms of absorption of caseinophosphopeptide-bound iron using this cell line as a tool.

The rat intestinal epithelial 2/4/A1 cell line forms polarized monolayers 4-6 days after seeding onto permeable supports. There are studies informing that the paracellular permeability of hydrophilic compounds is comparable to that of the small intestine, making this cell line a promising possible alternative to CACO-2 monolayers for studies of passive drug transport. However, an important disadvantage of 2/4/A1 cells must be noted: they are of rodent origin, and the regulation of gene expression during differentiation is probably not similar to that in humans [60]. However, this cell line can be useful to discriminate drugs transported by the paracellular route in the intestine. Moreover, the average pore diameter of the tight junctions in 2/4/A1 monolayers accurately reflects the average pore diameter of the human small intestine.

Intestinal absorption assays not involving intestinal cell lines, have also been developed. The Madin-Darby canine kidney (MDCK) cell line has received attention as an alternative to CACO-2 cells for permeability measurements. When grown under standard culture conditions, MDCK cells have been shown to differentiate into columnar epithelial cells with tight junctions close to the apical surface of the cells when cultured on semipermeable supports [61]. This provides a good *in vitro* model for transport at the intestinal level and has been demonstrated to be an adequate tool for assessing membrane permeability properties because of a good correlation between apparent permeability values and human absorption. The permeability of several highly and passively diffused drugs by MDCK cell lines correlates well with experiments performed using other cell cultures as *in vitro* intestinal absorption models [61]. However, in the parental MDCK cell line, no evidence for active transport of large neutral amino acids and bile acids was detected. The big advantage to using MDCK cells is the minimum culturing time, whereas, for example, CACO-2 cells need 21 days for the formation of a fully differentiated polarised cell monolayer. That could be the principal practical shortcoming of this cell line. Other non-intestinal cell lines such as LLc-PK1 cells are often used as models for epithelial tissue. For example, like the human intestinal epithelium, the porcine kidney epithelial cell line expresses microvilli and tight junctions [62]. CHO cells, derived from Chinese hamster ovary, allow researchers to characterize the membrane transport characteristics of several di- and tri-peptides. This cell line is used mainly for characterizing substrates and inhibitors of hPepT1, an apical transporter responsible for the uptake of a broad array of small peptides [63].

The literature reveals many studies on the intestinal absorption of compounds using human cell lines. However, in order to develop an adequate and precise study, it is very important to define the parameters for a cell-based *in vitro* permeability method (e.g., cell type, pH, transport direction, time and concentration of the assessed compound) and validate the method to support the formal biopharmaceutical classification system (BCS) of drugs. Twenty reference drugs were selected and permeability values determined using the Madin-Darby canine kidney type II cell line heterologously expressing the human P-glycoprotein transporter (MDCKII-MDR1). A rank order relationship was established between the *in vitro* permeability value and human intestinal absorption values. This relationship was as predicted and validated the MDCKII-MDR1 permeability method as defined by the BCS Guidance.

Cell culture models have been demonstrated to be useful for studying intestinal absorption and metabolism of carotenoids. The PF11 and TC7 clones of the CACO-2 cell line exhibit β-carotene-15,15′-oxygenase activity, a key enzyme in the conversion of carotenoids to vitamin A. An *in vitro* cell culture system used to study intestinal absorption of carotenoids [64] showed that under conditions mimicking the postprandial state, CACO-2 cells on membranes take up carotenoids and secrete them incorporated into chylomicrons. Both the cellular uptake and secretion of β-carotene are saturable, concentration-dependent processes. The selective absorption of all-trans β-carotene versus its cis isomers, the differential absorption of individual carotenoids and the specific interactions between carotenoids during their absorption are discussed. The participation of a specific epithelial transporter in the intestinal absorption of carotenoids is proposed [64].

A CACO-2 *in vitro* model appears to be suitable to investigate the transport of drugs across the intestinal epithelium; however, this model gives no information about the metabolic phase that follows the absorption of a drug but could provide information to investigate its pharmacokinetical behaviour [65]. A quick screen of multiple foods or food combinations is a useful tool for generating hypotheses for human bioavailability studies of a specific compound in absorption assays by cell lines [30].

In Vitro Cell Cultures as an Alternative to Animal Testing

Approximately 50 years ago, Russell and Burch launched for the first time the "Three Rs" concept about animal experimentation and the use of alternative methods to avoid the massive use of animals in research. This concept is based on Balls (2009) [66]:

- Replacement: the use of alternative methods that provide the same information as animals do
- Reduction: the minimal use of animals to obtain an objective proposed previously
- Refinement: the use of procedures that reduce the pain and suffering of animals

Since 1991, ECVAM (European Center for the Validation of Alternative Methods) has been responsible for validating alternative methods in order to make them the chosen protocol and to avoid animals being used unnecessarily [67]. There are several alternatives to animal testing, including human-based testing (mainly to study skin corrosion; e.g., EPISKIN and EPIDERM), computer simulations and cell cultures.

Cell cultures have been the most used to date, since samples of human cells and tissues can be used in different subjects (e.g., biochemistry and molecular biology, microbiology, toxicology and, of course, nutrition research). On one hand, the cost of cell cultures is relatively low, all conditions are well controlled in the laboratory atmosphere and the method provides researchers with information specifically relating to humans. On the other hand, using cell and tissue cultures does not allow a researcher to see the effect of a substance in a living body with all its complexities (absorption, transport, detoxification, etc.). Therefore, it can be concluded that alternative methods are valuable for research and can help limit the use of animals; however, they cannot replace them entirely. In nutrition research, animals are commonly used, but nowadays, some alternative methods have been validated to study certain aspects such as intestinal absorption. Some studies that have used *in vitro* models are exposed below.

The cell lines previously described in this chapter have been widely demonstrated to be useful in studying certain aspects of the metabolism in human enterocytes and the absorption and transport of different substances that arrive at the small intestine during the digestion of foods (e.g., amino acids, minerals, fatty acids, vitamins, or drugs). Moreover, the use of these cells has allowed researchers to elucidate the molecular mechanisms of the formation and assembly of lipoproteins; to obtain a rapid assessment of fatty acid, cholesterol and phospholipid uptake and metabolism; and to have the advantage of using human rather animal tissues [68]. The differences in biological characteristics (e.g., anatomy, metabolic rate) between animals and humans should be kept in mind while studying (i.e., questions devoted to identification of the diet component responsible for a specific function) [69]. The gut microenvironment is very complex, and there are many live organisms and cells interacting, such as bacteria (probiotic, commensal and pathogenic), absorptive cells, mucus-secreting cells and so on. However, nowadays, other intestinal cell lines proposed have been shown to be able to reproduce this environment in a reliable manner [70].

One of the inconveniences of *in vitro* models, such as cell cultures, is the extrapolation to animal models and subsequently to humans. As it has been mentioned before, the use of *in vitro* and *in vivo* models in nutrition go hand in hand. For example, in the report carried out by Hurley and McCormick (2003) [71], the infection mechanism of the intestinal epithelium induced by *Salmonella typhimurium* was studied using both *in vitro* and *in vivo* models. Tissue culture models have served as an extremely useful tool in evaluating the molecular mechanisms that underlie pathogenic disease processes. Investigators have been creative over the years in developing more complex *in vitro* models to better reconstruct the *in vivo* situation. Moreover, animal models allow evaluation of the biological significance of the effects recorded *in vitro* and testing of hypotheses on how a specific factor affects specific species under specific circumstances. Despite this, the leap from *in vitro* to *in vivo* models still introduces a plethora of variables that make interpretation extremely difficult. The use of *in vitro* models will continue and, through advances in cell and molecular biology (genomic and proteomic), should become more predictively accurate. As further knowledge is gained using a combination of both systems, a better understanding of disease pathogenesis, intestinal absorption mechanisms, gut microenvironmental conditions among other interesting aspects in nutrition research will emerge [71].

INTESTINAL CELL LINES AS AN *IN VITRO* MODEL OF CLINICAL VALUE

Crohn's disease and ulcerative colitis are chronic inflammatory bowel diseases (IBD) of unknown origin characterized by an infiltration of inflammatory cells into mucosal lesions. The main cells involved in the inflammatory response are monocytes/macrophages, polymorphonuclear leucocytes and endothelial cells, accompanied by the expression of a wide spectrum of pro-inflammatory cytokines and other mediators of inflammation that play an important role in the ethiopathogenesis of IBD [72]. Ultrastructural investigations have provided information about morphological and structural alterations of the intestinal barrier system in patients with IBD. These studies found that microvilli were reduced in number and length [73], and the tight junctions appeared to be abnormal. Thus, patients with IBD show altered absorption of different compounds at the intestinal level. Intestinal epithelial cells have a strategic position at the interface between the antigenic luminal environment and the internal milieu. Different studies have shown that epithelial cells actively contribute to the gut immune system, mediating processes such as mucosal defence and barrier repair [74], and playing a critical role in the inflammatory response. *In vitro* epithelial cells studies have played and are continuing to play a major role in providing specific information relevant to IBD, as their ability to synthesize and secrete inflammatory mediators themselves allows them to trigger immune cells and initiate the inflammatory process. Culture experiments have provided insights into the effects of individual cytokines and other inflammatory mediators, injury and repair, apoptosis, necrosis and other processes that may be involved in IBD [75]. A defective epithelial cell barrier functioning and exaggerated immune activity are thought to cooperate in a self-amplifying loop, where barrier dysfunction causes increased paracellular permeation of harmful luminal antigens and thus increases activation of mucosal immune cells. It has been found that inflammatory mediators can be released by various cell types including monocytes-macrophages, neutrophils and endothelial cells and has been found in increased amounts in the intestinal tissue of IBD patients [76]. Therefore, the choice of an adequate method, considering described aspects, to study inflammatory bowel disease seems to be crucial for characterization of the pathogenesis. *In vitro* models including intestinal cell lines could provide a valuable tool for experimental therapeutic research. In this regard, Tanoue et al. (2008) [72] and Van de Walle et al. (2010) [77] used the CACO-2 cell line to study inflammatory parameters at the intestinal level. An *in vitro* system for assessing the anti-inflammatory effect of food factors was developed by establishing a co-culture system with human intestinal epithelial CACO-2 cells and stimulated murine macrophage RAW264.7 cells. In this system, the study of the trans-epithelial electrical resistance, as a marker of the tight junction integrity (number and complexity of strands) of the CACO-2 monolayer, was used as a supplier of information on the mucosal injury induced by inflammation [72]. According to this study, the established gut inflammation *in vitro* model could imitate gut inflammation as seen in an IBD intestine. Van de Walle et al. (2010) [77] observed that the use of a cocktail of inflammatory stimuli at plausible concentrations on differentiated CACO-2 cells aimed at better reflecting the *in vivo* situation during on-going IBDs. This study also informs that this model could provide supplementary applications of the pathology at the intestinal epithelial level, since it could provide information on the release and interaction of inflammatory mediators as they occur in an inflammatory reaction. It could also provide data on the different cytokine profiles that are observed in the gut of Crohn's disease and ulcerative colitis patients, the anti-inflammatory mechanisms that aim at counteracting the inflammatory reaction and the involvement of additional cellular partners such as macrophages, dendritic cells or lymphocytes.

CONCLUSION

Any research approach in nutrition should be based upon the scientific method with the following steps: hypothesis formulation, experimentation under controlled conditions, results interpretation and theory formulation. Nutrition research is commonly focused on investigating food, composition, nutrient intake and associated biomarkers, nutritional utilization and nutrient functions as well as health-oriented aspects through epidemiological studies. The use of cell lines has facilitated the standardisation of many studies concerning nutrient functions and metabolism. Furthermore, cell lines allow the estimation of macronutrient synthesis and degradation, the changes in enzyme concentrations as well as the enzymatic activities of key molecules, providing valuable data concerning nutrient roles and utilization at the tissue level by assessing the involvement of specific nutrients in body homeostasis.

In molecular biology, cell cultures provide powerful tools to evaluate and establish metabolic pathways and regulatory roles of nutrients. The influence of some nutrients on cell physiology can be estimated through RNA and DNA values. The use of *in vitro* and *in vivo* models is likely to continue, as the effects of nutrition on health and disease cannot be fully explained without understanding nutrient actions at the nuclear level and their role in intra- and intercellular signal transduction (Mortensen et al., 2008). At the sub-cellular level (mitochondria, nuclei, etc.), it can characterize individual functions of different minerals, vitamins, fatty acids and so on, since some dietary constituents can govern the expression of proteins (i.e., carbohydrates and polyunsaturated fatty acids may control the cytosolic content of specific transcripts by regulating the processing events involved in mRNA maturations). In addition, certain nutrient functions can be elucidated at the gene level. Transfection studies involve the insertion of DNA into a cell to examine nutrient function and metabolism. Gene regulation by nutrients has been assessed in different isolated cells and tissues using appropriate markers of gene expression. Based on the information reported, deeper studies are necessary regarding the use of cell lines in nutrition research (and specifically in nutrient absorption at the intestinal level). These studies will help to reproduce physiological conditions with accuracy and to obtain an adequate correlation between *in vivo* and *in vitro* studies. Further information is needed not only on choosing an adequate type of cell line but also on the physiological conditions implicated.

ACKNOWLEDGMENTS

To the Fundaciòn Seneca, Agencia Regional de Ciencia y Tecnologìa, Regiòn de Murcia for the postdoctoral fellowship of C. Frontela, 09292/PD/08 and for the project 11978/PI/09

REFERENCES

[1] Fleischauer AT., Arab L. *Garlic and cancer: a critical review of the epidemiologic literature.* Nutr Cancer. 2003;46(1):15-22.

[2] Le Ferrec E, Chesne C, Artusson P, Brayden D, Fabre G, Gires P, Guillou F, Rousset M, Rubas W, Scarino ML. In vitro models of the intestinal barrier. The report and recommendations of ECVAM Workshop 46. European Centre for the Validation of Alternative methods. Altern Lab Anim, (2001);29: 649-68.

[3] Wikman A, Karlsson J, Carlstedt I, Artursson P. *A drug absorption model based on the mucus layer producing human intestinal goblet cell line HT29-H.* Pharm Res. 1993 Jun;10(6):843-52.

[4] Walter E, Janich S, Roessler BJ, Hilfinger JM, Amidon GL. *HT29-MTX/CACO-2 cocultures as an in vitro model for the intestinal epithelium: in vitro-in vivo correlation with permeability data from rats and humans.* J Pharm Sci. 1996 Oct;85(10):1070-6.
[5] Mueller-Klieser W. *Three-dimensional cell cultures: from molecular mechanisms to clinical applications.* Am J Physiol. 1997 Oct;273(4 Pt 1):C1109-23.
[6] Miura Y., Shiomi H., Sakai F., Yagasaki K. *Assay systems for screening food components that have anti-proliferative and anti-invasive activity to rat ascites hepatoma cells: In vitro and ex vivo effects of green tea extract.* Cytotechnology, (1997);23:127-132.
[7] Kim YS., Milner J.A. *Bioactive Food Components and Cancer-Specific Metabonomic Profiles.* J Biomed Biotech. 2010;(2011), Article ID 721213, doi:10.1155/2011/721213.
[8] Mortensen A., Sorensen IK., Wilde C., Drogani S., Mullerova D., Toussaint O., Zloch Z., Sgaragli G., Ovesna J. *Biological models for phytochemical research: from cell to human organism.* Br J Nutr. 2008;99:118-126.
[9] Garrett DA., Failla ML., Sarama RJ. *Development of an in vitro digestion method to assess carotenoid bioavailability from meals.* J Agric Food Chem 1999;47:4301–4309.
[10] Yun S., Habicht JP., Miller DD., Glahn RP. *An in vitro digestion/CACO-2 cell culture system accurately predicts the effects of ascorbic acid and polyphenolic compounds on iron bioavailability in humans.* J Nutr 2004; 134:2717-2721.
[11] Yamashita S., Furubayashi T., Kataoka M., T. Sakane, Sezaki H., Tokuda H. *Optimized conditions for prediction of intestinal drug permeability using CACO-2 cells.* European Journal of Pharmaceutical Sciences, 2000;10(3):195-204.
[12] Tsuji A, Takanaga H, Tamai I, Terasaki T.*Transcellular transport of benzoic acid across CACO-2 cells by a pH-dependent and carrier-mediated transport mechanism.* Pharm Res. 1994;11(1):30-7.
[13] Artursson P, Karlsson J. *Correlation between oral drug absorption in humans and apparent drug permeability coefficients in human intestinal epithelial (CACO-2) cells.* Biochem Biophys Res Commun. 1991;175(3):880-5.
[14] Varma MV, Panchagnula R. *pH-dependent functional activity of P-glycoprotein in limiting intestinal absorption of protic drugs: kinetic analysis of quinidine efflux in situ.* J Pharm Sci. 2005;94(12):2632-43.
[15] Verwei M., Freidig AP., Havenaar R., Groten JP. *Predicted Serum Folate Concentrations Based on In Vitro Studies and Kinetic Modeling are Consistent with Measured Folate Concentrations in Humans.* J. Nutr. 2006;136:3074-3078.
[16] Frazer and Anderson. Iron Imports. I. *Intestinal iron absorption and its regulation.* Am J Physiol Gastrointest Liver Physiol 2005; 289:631–635.
[17] Mahler GJ, Shuler ML, Glahn RP. *Characterization of CACO-2 and HT29-MTX cocultures in an in vitro digestion/cell culture model used to predict iron bioavailability.* J Nutr Biochem. 2009;20(7):494-502.
[18] Boyer J., Hai Liu R. *Apple phytochemicals and their health benefits.*Nutrition Journal 2004; 3:5.
[19] Bohnsack BL, Hirschi KK. *Nutrient regulation of cell cycle progression.* Annu Rev Nutr. 2004;24:433-453, 2004.
[20] Yamaji S, Tennant J, Tandy S, Williams M, Singh Srai SK, Sharp P. *Zinc regulates the function and expression of the iron transporters DMT1 and IREG1 in human intestinal CACO-2 cells.* FEBS Lett 2001; 507: 137-41.

[21] Reboul E, Thap S, Tourniaire F, André M, Juhel C, Morange S, Amiot Carlin MJ, Lairon D & Borel P (2007b) *Differential effect of dietary antioxidant classes (carotenoids, polyphenols, vitamin C and vitamin E) on lutein absorption.* Br J Nutr 97, 440-446.

[22] Pinto, M., Rabine-Leon, S., Appay, M.D., Kedinger, M., Triadou, N., Dussaulx, E., Louroix, B., Simon-Assmann, P. and Haffeb, K. *Enterocyte-like differentiation and polarisation of the human colon carcinoma cell line CACO-2 in culture.* Biology of the Cell 1983;47: 323–330.

[23] Skolnik S, Lin X, Wang J, Chen XH, He T, Zhang B. Towards prediction of in vivo intestinal absorption using a 96-well CACO-2 assay. J Pharm Sci. 2010;99(7):3246-65.

[24] Heikkinen AT, Mönkkönen J, Korjamo T. *Kinetics of cellular retention during CACO-2 permeation experiments: role of lysosomal sequestration and impact on permeability estimates.* J Pharmacol Exp Ther. 2009;328(3):882-92.

[25] O' Sullivan L, Ryan L, Aherne SA, O' Brien NM. *Cellular transport of lutein is greater from uncooked rather than cooked spinach irrespective of whether it is fresh, frozen, or canned.* Nutr Res 2008;28(8):532-8.

[26] Laparra JM, Glahn RP, Miller DD. *Assessing potential effects of inulin and probiotic bacteria on Fe availability from common beans (Phaseolus vulgaris L.) to CACO-2 cells.* J Food Sci. 2009;74(2):H40-H46.

[27] Lonnerdal B. *Soybean ferritin: implications of iron status of vegetarians.* Am J Clin Nutr 2009;89(5):1680-1685.

[28] Stadler NC, Somoza V, Schwab W (2009). *Absorption of 3(2H)-furanones by human intestinal epithelial CACO-2 cells.* J Agric Food Chem 2009; 57 (9):3949–3954.

[29] Wang Z., Clifford MN., Sharp P. *Analysis of chlorogenic acids in beverages prepared from Chinese health foods and investigation, in vitro, of effects on glucose absorption in cultured CACO-2 cells.* Food Chemistry 2008;108(1):369-373.

[30] Frontela C, Scarino ML, Ferruzza S, Ros G, Martínez C. (2009) Effect of dephytinization on bioavailability of iron, calcium and zinc from infant cereals assessed in the CACO-2 cell model. World J Gastroenterol 2009;15:1977-84.

[31] Hilgendorf C., Spahn-Langguth H., Regardh CG., Lipka E., *Amidon GL., P. Langguth P. CACO-2 versus CACO-2/HT29-MTX cocultured cell lines: permeabilities via diffusion, inside- and outside-directed carrier mediated transport.* J. Pharm. Sci. 2000;89:63-75.

[32] Lown KS., Bailey DG., Fontana RJ., Janardan SK., Adair CH., Fortlage LA, Brown MB, Guo W., Watkins PB. *Grapefruit Juice Increases Felodipine Oral Availability in Humans by Decreasing Intestinal CYP3A Protein Expression.* J Clin Invest 1997;99(10):2545–2553.

[33] Tetsuya A., Mitsuru S., Shiro F., Yukiya H. The effects of culture conditions on CYP3A4 and MDR1 mRNA induction by 1.alpha.,25-dihydroxyvitamin D3 in human intestinal cell lines, CACO-2 and LS180. drug metab pharmacokinet, 2005;20, 4: 268-274.

[34] Crespi CL, Penman BW, Hu M. *Development of CACO-2 cells expressing high levels of cDNA-derived cytochrome P4503A4.* Pharm Res 1996;13:1635–1641.

[35] Engman HA., Lennernäs H., Taipalensuu J., Otter C., Leidvik B., Artursson P. *CYP3A4, CYP3A5, and MDR1 in human small and large intestinal cell lines suitable*

for drug transport studies. Journal of Pharmaceutical Sciences. 2001;90(11):1736–1751.

[36] Pontier C., Pachot J., Botham R., Lanfant B., Arnaud P. *HT29-MTX and CACO-2/TC7 monolayer as predictive models for human intestinal absorption: role of the mucus layer.* J Pharm Sci 2001; 90(10):1608-1619.

[37] Boyer J., Brown D., Liu RH. *Uptake of quercetin and quercetin 3-glucoside from whole onion and apple peel extracts by CACO-2 cell monolayers.* J Agric Food Chem., 2004;52:7172-7179.

[38] Fairweather-Tait S, Lynch S, Hotz C, Hurrell RF, Abrahamse L, Beebe S, Bering S, Bukhave K, Glahn R. *The usefulness of in vitro models to predict the bioavailability of iron and zinc - A consensus statement from the HarvestPlus expert consultation.* Int J Vitam Nutr Res 2005;75:371–4.

[39] Johnson, D.M., Yamaji, S., Tennant, J., Srai, S.K., Sharp, P.A. *Regulation of divalent metal transporter expression in human intestinal epithelial cells following exposure to non-haem iron.* FEBS Letters. 2005;579(9):1923-9.

[40] Mahler GJ., Shuler ML., Glahn RP. *Characterization of CACO-2 and HT29-MTX cocultures in an in vitro digestion/cell culture model used to predict iron bioavailability.* The Journal of Nutritional Biochemistry 2009;20:494–502.

[41] Hu M., Ling J., Lin H., Chen J. *Use of CACO-2 Cell Monolayers to Study Drug Absorption and Metabolism* Optimization in Drug Discovery: In Vitro Methods *Series:* Methods in Pharmacology and Toxicology 2004. DOI: 10.1385/1-59259-800-5:019.

[42] Turco L, Catone T, Caloni F, Consiglio ED, Testai E, Stammati A. *CACO-2/tc7 cell line characterization for intestinal absorption: How reliable is this in vitro model for the prediction of the oral dose fraction absorbed in human?* Toxicol In Vitro 2010;20.

[43] Soler, A.; Romero, M. P.; Macia, A.; Saha, S.; Furniss, C. S. M.; Kroon, P. A.; Motilva, M. J. *Digestion stability and evaluation of the metabolism and transport of olive oil phenols in the human small-intestinal epityhelial CACO-2/TC7 cell line* Food Chem. 2010;119:703–714.

[44] Zucco F., Batto AF., Bises G., Chambaz J., Chiusolo A., Consalvo R., Cross H., Dal Negro G., De Angelis I., Fabre G., Guillou F., Hoffmann S., Laplanche L., Morel E., Pinçon-Raymond M., Prieto P., Turco L., Ranaldi G., Rousset M., Sambuy Y., Scarino ML., Torreilles F., Stammat A. *An inter-laboratory study to evaluate the effects of medium composition on differentiation and barrier function of CACO-2 cell lines,* ATLA 2005;33:603–618.

[45] Kirkland SC. *Dome formation by a human colonic adenocarcinoma cell line (HCA-7).* Cancer Res., 1985;45:3790-3795.

[46] Buck S., Buttery LD., Hawkey C., Rose FRAJ. *Development of an in vivo-like in vitro Intestinal Epithelial Model.* European Cells and Materials 2006; 11(3):67.

[47] Zweibaum, A., Laburthe, M., Grasset, E., Louvard, D. *Use of cultured cell lines in studies of intestinal cell differentiation and function.* In Handbook of Physiology. The Gastrointestinal System. Vol IV:Intestinal Absorption and Secretion (ed. M. Field and R. A. Frizzel), Amer. Physiol. Soc. 1991;223-255.

[48] Baricault L., Denariaz G., Houri JJ., Bouley C., Sapin C., Trugnan G. *Use of HT-29, a cultured human colon cancer cell line, to study the effect of fermented milks on colon cancer cell growth and differentiation. Carcinogenesis* 1995;16(2): 245-252.

[49] Story E., Harris G. *Effects of green tea and EGCC on IL-8 inflammatory responses in LPS and TNF-α induced differentiated HT-29 hu8man colon cancer cells.* FASEB J 2010;24 (meeting abstract suppl.)541.5.

[50] Cosentino S., Gravaghi C., Donetti E., Donida BM., Lombardi G., Bedoni M., Fiorilli A., Tettamanti G., Ferraretto A. *Caseinphospho peptide-induced calcium uptake in human intestinal cell lines HT-29 and CACO-2 is correlated to cellular differentiation.* The Journal of Nutritional Biochemistry. (2010), 21(3): 247-254.

[51] Nakamura T, Sakaeda T, Ohmoto N, Moriya Y, Komoto C, Shirakawa T, Gotoh A, Matsuo M, Okmura K.. Pharm Res 2003; 20(2):324-7. *Gene expression profiles of ABC transporters and cytochrome P450 3A in CACO-2 and human colorectal cancer cell lines.*

[52] Said HM, Ortiz A, Moyer MP, Yanagawa N. (2000) Riboflavin uptake by human-derived colonic epithelial NCM460 cells. Am J Physiol Cell Physiol. 2000 Feb;278(2):C270-6.

[53] Said HM, Ortiz A, Subramanian VS, Neufeld EJ, Moyer MP, Dudeja PK. (2001) Mechanism of thiamine uptake by human colonocytes: studies with cultured colonic epithelial cell line NCM460. Am J Physiol Gastrointest Liver Physiol. 2001 Jul;281(1):G144-50.

[54] van Klinken BJ, Oussoren E, Weenink JJ, Strous GJ, Büller HA, Dekker J, Einerhand AW. (1996) The human intestinal cell lines CACO-2 and LS174T as models to study cell-type specific mucin expression. Glycoconj J. 1996 Oct;13(5):757-68.

[55] López-Posadas R, Ballester I, Mascaraque C, Suárez MD, Zarzuelo A, Martínez-Augustin O, Sánchez de Medina F. (2010) Flavonoids exert distinct modulatory actions on cyclooxygenase 2 and NF-kappaB in an intestinal epithelial cell line (IEC18). Br J Pharmacol. 2010 Aug;160(7):1714-26.

[56] Duizer E, Gilde AJ, Versantvoort CH, Groten JP. (1999) Effects of cadmium chloride on the paracellular barrier function of intestinal epithelial cell lines. *Toxicol* Appl Pharmacol. 1999 Mar 1;155(2):117-26.

[57] Thomas C, Oates PS. (2002) IEC-6 cells are an appropriate model of intestinal iron absorption in rats. J Nutr. 2002 Apr;132(4):680-7.

[58] Rajgopal A, Sierra EE, Zhao R, Goldman ID. Expression of the reduced folate carrier SLC19A1 in IEC-6 cells results in two distinct transport activities. Am J Physiol Cell Physiol. 2001 Nov;281(5):C1579-86.

[59] Pérès JM, Bouhallab S, Bureau F, Neuville D, Maubois JL, Devroede G, Arhan P, Bouglé D. Mechanisms of absorption of caseino phosphor peptide bound iron. J Nutr Biochem. 1999 Apr;10(4):215-22.

[60] Perreault N, Beaulieu JF. Primary cultures of fully differentiated and pure human intestinal epithelial cells. Exp Cell Res. 1998 Nov 25;245(1):34-42.

[61] Irvine JD, Takahashi L, Lockhart K, Cheong J, Tolan JW, Selick HE, Grove JR. MDCK (Madin-Darby canine kidney) cells: A tool for membrane permeability screening. J Pharm Sci. 1999 Jan;88(1):28-33.

[62] Chang JH, Benet LZ. Glucuronidation and the transport of the glucuronide metabolites in LLC-PK1 cells. Mol Pharm. 2005 Sep-Oct;2(5):428-34.

[63] Fujisawa Y, Tateoka R, Nara T, Kamo N, Taira T, Miyauchi S. The extracellular pH dependency of transport activity by human oligopeptide transporter 1 (hPEPT1)

expressed stably in Chinese hamster ovary (CHO) cells: a reason for the bell-shaped activity versus pH. Biol Pharm Bull. 2006 May;29(5):997-1005.

[64] During A., Harrison EH. *An in vitro model to study the intestinal absorption of carotenoids.* Food Research International Volume 38, Issues 8-9, October-November 2005, Pages 1001-1008.

[65] Rossi, A., Poverini, R., Di, Lullo G., Modesti, A., Modica, A. and Scarino, M.L. *Heavy metal toxicity following apical and basolateral exposure in the human intestinal cell line Caco2.* Toxicology in Vitro 1996; 10: 27–36.

[66] Balls M. *The origins and early days of the Three Rs concept.* Altern Lab Anim.2009; 37: 255-65.

[67] Spielmann H. (2003) *The future of animal experiments in nutrition research.* Forum Nutr.; 56: 297-9.

[68] Levy E, Mehran M, Seidman E. (1995) CACO-2 cells as a model for intestinal lipoprotein synthesis and secretion. FASEB J.; 9: 626-35. Review.

[69] Mortensen A., Sorensen IK., Wilde C., Dragoni S., Mullerova D., Toussaint O., Zloch Z., Sgragli G., Ovesna J. *Biological models for phytochemical research: from cell to human organism.* Br J Nutr 2008;99:118-126.

[70] Kemperman R. A., Bolca S., Roger L. C. , Vaughan E. E. *Novel approaches for analysing gut microbes and dietary polyphenols: challenges and opportunities.* Microbiology (2010), 156, 3224–3231.

[71] Hurley BP and McCormick BA. (2003) *Translating tissue culture results into animal models: the case of Salmonella typhimurium.* Trends Microbiol.; 11: 562-9.

[72] Tanoue T, Nishitani Y, Kanazawa K, Hashimoto T, Mizuno M. In vitro model to estimate gut inflammation using co-cultured CACO-2 and RAW264.7 cells. Biochem Biophys Res Commun. 2008 Sep 26;374(3):565-9. Epub 2008 Jul 21.

[73] Dvorak AM., Dickersin GR (1980). *Crohn's disease: transmission electron microscopic studies. Barrier function. Possible changes related to alterations of cell coat, mucous coat, epithelial cells, and paneth cells.* Human pathol 11:561-571.

[74] Cario E, Rosenberg IM, Brandwein SL, Beck PL, Reinecker HC, Podolsky DK..Lipopolysaccharide activates distinct signaling pathways in intestinal epithelial cell lines expressing Toll-like receptors. J Immunol. 2000;164(2):966-72.

[75] McKay DM, Philpott DJ, Perdue MH. Review article: In vitro models in inflammatory bowel disease research--a critical review. Aliment Pharmacol Ther. 1997 Dec;11 Suppl 3:70-80. Review.

[76] Reimund JM, Wittersheim C, Dumont S, Muller CD, Baumann R, Poindron P, Duclos B. Mucosal inflammatory cytokine production by intestinal biopsies in patients with ulcerative colitis and Crohn's disease. J Clin Immunol. 1996 May;16(3):144-50.

[77] Van De Walle J, Hendrickx A, Romier B, Larondelle Y, Schneider YJ. Inflammatory parameters in CACO-2 cells: effect of stimuli nature, concentration, combination and cell differentiation. Toxicol In Vitro. 2010;24(5):1441-9.

In: CACO-2 Cells and their Uses
Editor: Megan A. Schulz, pp. 49-60
ISBN 978-61209-917-0

Chapter 3

CO-CULTURE SYSTEMS WITH CACO-2 CELLS TO EVALUATE THE BIOACTIVITY OF PLANT EXTRACTS

Montserrat Pinent, M. Teresa Blay and Anna Ardévol*
Department of Biochemistry and Biotechnology.
Universitat Rovira i Virgili. Marcel·lí Domingo s/n.
43007 Tarragona, Spain

ABSTRACT

Natural extracts are potential candidates for the development of new functional foods, but these extracts are usually complex mixtures of molecules of uncertain bioavailability and are often partially metabolized before reaching the target cells *in vivo*. Therefore, *in vitro* studies of the bioactivity of these extracts in cell culture may not translate to real-life applications *in vivo*. To overcome this limitation, co-culture systems that simulate the intestinal barrier using CACO-2 cells have been developed. Human CACO-2 enterocytes seeded onto culture inserts can be co-cultured with the cell type in which the plant extract effects are to be analyzed. Using this system, plant extracts applied to the apical side of CACO-2 cells have to be absorbed and metabolized by the cell barrier to reach the cells of interest. The system has been tested with different cell types such as pancreatic rat beta cells (INS-1) and human hepatocytes (HEPG2). CACO-2 cells in co-culture maintain their abilities to filter and metabolize the tested plant extracts. In the same way, key functions of the cell type of interest (i.e. triglyceride secretion in hepatocytes and glucose-stimulated insulin secretion in INS-1) are not affected by the co-culture. Therefore co-culture of CACO-2 cells with other cell types seems to be a suitable method to evaluate the bioactivity of natural plant extracts. However, there are some aspects that must be carefully considered before generalizing the appropriateness of this method. First, the co-culture may induce some adaptations in one or both cell types. For example, in INS-1 and HEPG2 co-culture systems, gene expression of the glucose transporter SGLT-1 is modified. Thus it should be considered whether adaptations may affect the function of interest and tested for. Furthermore, long-term co-culture can influence the viability of cells. Therefore, the time frame in which the

*Corresponding author and requests for reprints: Anna Ardévol, Departament de Bioquímica i Biotecnologia, C. Marcel.lí Domingo, s/n, 43007, Tarragona, Spain., Phone number: 34 977 559566, Fax number: 34 977 558232, E-mail: anna.ardevol@urv.cat

cells are normally functional must be established for each cell type. As an example, HEPG-2 cells can be co-cultured for at least 24 hours without effects on cell viability while INS-1 co-culture studies should be restricted to shorter times (i.e. 4 hours).

INTRODUCTION

The physiological effects of functional foods can only be demonstrated after numerous studies. The bioactive compound must first be identified before final intervention studies on humans can be conducted to provide the human data required to make a nutritional/health claim regarding a functional food [1]. This often involves screening a large number of molecules. As animal experimentation should be restricted as much as possible, it is advisable to initially screen bioactive compounds in cell lines (European Union).

Extracts from natural sources are often used for developing new functional foods. However, these extracts are complex mixtures of molecules that have to be absorbed and metabolized if they are to finally reach the cells *in vivo* [2,3,4]. Therefore, directly applying extracts to cell cultures may not faithfully reflect the *in vivo* conditions [5]. To overcome this problem, many studies have analyzed the activity of pure molecules [6-10]. However, obtaining pure components may prove difficult because of the technical limitations linked to the great diversity of molecules in the extracts and the low quantity in which they are found. In addition, few pure molecules have been detected in plasma, and those that have are usually modified forms of the original ingested molecule [3,11,12]. Therefore, working with pure molecules directly to analyze their bioactivity in cell lines does not completely overcome the problem of whether a potentially beneficial compound will be absorbed. Finally, and very importantly, the *in vivo* effects of natural extracts may be due not only to the pure molecules but also to synergistic interactions between several molecules [13].

The absorption of extract compounds implies crossing the intestinal epithelium barrier, which can modify some absorbed compounds. To study absorption, several *in vitro* models have been developed, with the CACO-2 cell line being one of the most studied [14,15]. Human intestinal epithelial CACO-2 cells grown on permeable inserts are able to differentiate and polarize, forming a continuous monolayer with tight junctions that mimics the intestinal barrier [16,17]. These cells thus attain many of the morphological and functional characteristics of intestinal enterocytes and enable intestinal permeability to be evaluated [18]. CACO-2 cell monolayers have been widely accepted as an *in vitro* model membrane for screening intestinal absorption of drugs [14,15,19]. Furthermore, they have been used as an *in vitro* model to study the absorption of natural bioactive compounds [8,15,18,20]. Therefore, to study the effects of bioactive compounds on target cells in a more physiological manner, a system involving a co-culture of intestinal epithelial CACO-2 cells, which physically separates the desired target cells from the bioactive compound, would be useful. Remarkably, the co-culture method has been shown to reproduce situations that resemble the *in vivo* coexistence of several cell types and their interactions; e.g., adipocyte-macrophage cross-talk [21] and the improvement in gastrointestinal permeability modeling [22]. Also, the CACO-2 co-culture system has been used to simulate the intestinal barrier and has proven to be a better system than single-layered cells to analyze toxicity in target cells such as human fibroblast TIG-1 cells [23] and human hepatocytes (HEPG2) [24].

The goal of this chapter is to describe and evaluate the viability of co-culture of target cells under a monolayer of CACO-2 cells to simulate the intestinal barrier, with which the bioactivity of natural extracts can be tested. This method has been tested in hepatocytes (human cell line HEPG2) and pancreatic beta cells (rat cell line INS-1 and mouse cell line MIN-6). We illustrate the advantages and limitations of such co-culture systems by considering key points of the method, such as viability of the co-culture, possible modulation of the action of CACO-2 cells as a barrier for natural extracts by the fact that they are in co-culture, and effectiveness of the absorbed bioactive compounds on the target cells.

CO-CULTURE SYSTEMS USING CACO-2 CELLS: CELL VIABILITY AND FUNCTIONALITY

The methodology based on the co-culture of target cells under CACO-2 cells has been tested and optimized with different cells types, including hepatocytes (human cell line HEPG2) and pancreatic beta cells (rat cell line INS-1 and mouse cell line MIN-6). HEPG2 cells have been used because many natural extracts show bioactivity in this cell type, as well as in the liver [25-28]. Therefore the established effects can serve to test the co-culture system. In addition, the key role of the liver in metabolism highlights the interest of this co-culture. On the other hand, the bioactivity of plant extracts on β-pancreatic cells has received little attention [29], and there is in fact much controversy, at least regarding flavonoid compounds, about the effects of plant extracts on β-cells. Therefore, it was of interest to develop a more physiological method, which would help in understanding such effects.

In co-culture systems that are used to analyze the effects of bioactive compounds on target cells, CACO-2 cells are seeded onto hanging semipermeable cell culture inserts and allowed to differentiate for 21 days, forming a confluent monolayer. The inserts with the differentiated CACO-2 cells are then placed onto the target cells, which have been seeded on cell culture plates (Figure 1).

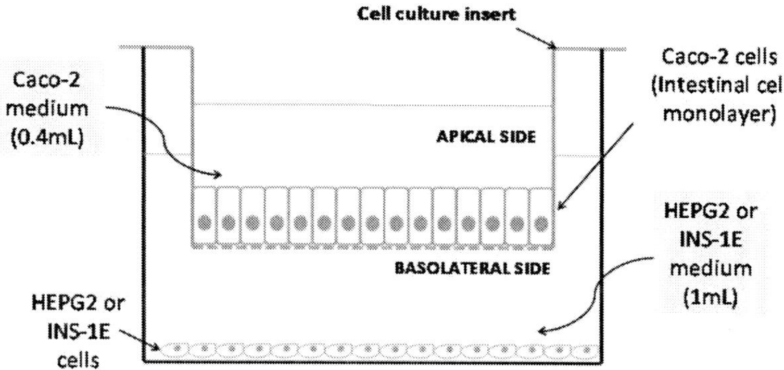

Figure 1. Scheme of the double-layered system. CACO-2 cells were seeded on inserts at a cell density of 5.3·104 cells/cm2. Co-culture was performed either with HEPG2 or INS-1E cells. HEPG2 cells were seeded onto 12-well plates at a cell density of 1·105 cells/cm2. At day 20 for CACO-2 and day 3 for HEPG2 cells, both cell types were co-cultured. INS-1E cells were seeded onto 12-well plates at a cell density of 1·105 cells/cm2; after the cells reached confluency, they were co-cultured for 4 and 24 hours.

This methodology has been used to co-culture CACO-2 and HEPG2 cells [24,30]. Co-culture durations of up to five days did not alter CACO-2 barrier permeability or functionality, as assessed by measuring the TEER value. TEER values were constant, did not differ from those of CACO-2 cells grown alone, and were similar to published values [30,31]. No change in the TEER value suggested no changes in leakiness or tightness in the paracellular pathway [30], which is used by some compounds to cross the intestinal barrier. Similarly, the co-culture of CACO-2 with INS-1 cells up to 24 hours or MIN-6 cells up to five days does not modify the permeability barrier [32].

To better characterize the effects of co-culture on CACO-2 cells, the molecular marker sodium/glucose cotransporter 1 (SGLT1) was monitored, which is found in the apical side of CACO-2 cells, similarly to human jejunum absorptive cells [33]. Co-culture with HEPG2 cells leads to a down-regulation of SGLT1 gene expression in CACO-2 cells [30]. Conversely, SGLT1 gene expression is induced by co-culture with pancreatic-beta cells [32]. In these experiments, the changes in SGLT1 gene expression did not lead to changes in the concentration of glucose that reached the basolateral side. However, SGLT1 seems to act as a transporter for some flavonoid glucosides. In particular, quercetin glucosides have been shown to use SGLT1 [34]. Although there is controversy regarding such transport [35], the decrease in SGLT1 should be considered when analyzing extracts enriched in these flavonoids. However, it should not influence experiments performed with extracts containing other flavonoids that might not be transported by SGLT1.

The co-culture systems also caused adaptive changes to the cells grown on the basolateral side [30]. The co-culture system affected the morphology of HEPG2 cells, which showed a slight tendency to be elongated. Furthermore, expression of glucose-6-phosphatase was down-regulated, and regulation by stimuli such as AMPc and insulin was less pronounced than in non-co-cultured hepatocytes [30]. The effects on lipid metabolism in HEPG2 cells as a result of co-culture have also been investigated. After six hours of incubation with radioactive acetate, the amount of cholesterol and triglycerides synthesized was lower in co-cultured than in non co-cultured hepatocytes, while the amount of cholesterol and triglycerides secreted into the cell culture media was similar in both culture situations [30]. However, cell function is not always attenuated by co-culture. Choi et al. [24] found that the activity of the enzyme CYP1A172 induced with a typical inducer (3MC) in CACO-2 and in HEPG-2 cells was slightly, though not significantly, enhanced in a co-culture system compared to a pure culture system. Thus, co-culture of CACO-2 cells with other cells might lead to different functional responses than pure culture, with these differences being dependent on the function to be analyzed.

Concerning beta-cells in co-culture, morphology was also affected by co-culture, with a phenotype more typical of an undifferentiated cell. Viability assays in INS-1 cells showed no damage after a short-term co-culture (four hours). Beta-cell function assessed by glucose-stimulated insulin secretion (GSIS) was maintained in this system, with levels equivalent to pure cultures (two-fold stimulation). However, longer periods (24 hours) of co-culture significantly reduced cell viability [32]. Considering these results, when using a co-culture system, the effects on cell viability must be carefully considered.

With the previous considerations in mind, a two-day co-culture of CACO-2 cells seeded onto culture inserts over a monolayer of human macrophage-like THP cells caused apoptotic and necrotic cell death in the CACO-2 cells. In this case, the co-culture was used as a model for intestinal inflammation because the differentiated THP cells secrete cytokines responsible for toxic effects, such as TNFα [36]. Other inflammation models involving the co-culture of CACO-2 cells over a monolayer of macrophages have been proposed, although they are beyond the goal of this chapter [37].

Taken together, these studies suggest that cells in co-culture might behave slightly differently than single-layered cultures. While co-culture could better resemble a physiological context, it is important to determine whether co-culture-induced adaptations could compromise the objective of the study.

WORKING WITH NATURAL EXTRACTS: CACO-2 AS A FILTER TO SIMULATE THE HUMAN INTESTINAL BARRIER

The interest of using co-culture systems to analyze the effects of natural extracts depends on the capacity of the CACO-2 monolayer to act as an intestinal barrier. Thus, only the absorbed components of the extract will reach the target cells. The intestinal barrier is also an active modifier of the components of the extracts. Natural extracts contain a variety of compounds with potentially beneficial effects such as fiber, vitamins, phytosterols, sulfur compounds, carotenoids, organic acids and polyphenols. Polyphenols are increasingly recognized as having beneficial effects on human health, and we are therefore going to focus this chapter on these substances, although the use of CACO-2-based co-cultures could be extended to study the effects of other compounds in natural extracts. In addition, this chapter discusses matrix effects that might affect absorption and metabolism. Polyphenols include a variety of compounds divided in several classes according to their chemical structure. Because they are extensively altered during first-phase metabolism, the molecular forms that reach the peripheral circulation and tissues are often different from those in foods [3]. Their metabolism can include several processes such as deglycosylation of the glycoside forms (the forms which are predominantly found in plants), glucuronidation, sulfatation, methylation and galloylation; the bioavailability of polyphenols varies among the different groups [3], [38]. Furthermore, intestinal microbiota are also responsible for the metabolism of some polyphenols, which are degraded into aromatic acids by the microflora [3]. This last point cannot currently be resolved with the co-culture model of CACO-2 cells; concerning metabolism, however, CACO-2 cells possess the enzymes that are responsible for glucuronidation and sulfation [39,40].

Furthermore, CACO-2 cells are of human origin, and several studies have demonstrated coincident results between data of bioactive compounds found in CACO-2 monolayer systems and in human plasma, supporting the notion that these cells are a good model for simulating polyphenol absorption in human intestines. However, caution must be taken when comparing results with those obtained in rodent models, which could have a different metabolism. To address this question we will first focus on studies performed with pure monomers, whose absorption and metabolism can be more easily analyzed than those of polyphenol mixtures as found in natural extracts, which can include polymerized forms.

Several studies have shown that CACO-2 cells are able to metabolize pure monomers. The flavonoid chrysin loaded onto the apical side of CACO-2 cells grown on culture inserts can be found on the basolateral side in an extensively glucuronidated and sulfated form [39-41]. Similarly, chrysin conjugates are also found in plasma in humans [42]. Treatment of a monolayer of CACO-2 cells with the tea flavonoid epigallocatechin gallate (EGCG) results in low biovailability, which appears on the basolateral side unchanged, similar to that found in humans after oral EGCG administration [43]. On the other hand, other tea catechins such as epicatechin and epigallocatechin are metabolized by CACO-2 cells, with the major forms being sulfated and methylated, which is consistent with human data that show the same forms in plasma [44]. Different data is obtained in rats; after administration of epicatechin, glucuronide was the major metabolite [10,45], suggesting metabolic differences between species. A study with the flavanone hesperidin also showed that metabolite formation was similar in a CACO-2 monolayer and in human plasma [46]. Differences were also observed in the absorption of the flavonoid luteolin between rat and human. Free luteolin and glucuronide and sulfate conjugates have been found in the plasma of rats administered with luteolin [47] while in human plasma, phase II metabolites (sulfates and glucuronides), but no free luteolin, have been found after ingestion of artichoke leaf extract [48]. Also in CACO-2 cell monolayers, luteolin treatment led to no free luteolin on the basolateral side, but to glucuronidates and sulfated forms [49]. Interestingly, this paper showed that luteolin is better absorbed and more efficiently metabolized when administered as a plant extract (*Artemisia afra*, a medicinal plant) than as pure solutions of both luteolin forms, suggesting that the plant matrix may facilitate uptake and metabolism. Interference of the food matrix with compound absorption *in vivo* [50] and in CACO-2 cells [51] has been described, although the results vary according to the tested compound and experimental conditions [52]. One matrix element that could interfere with bioavailability is the presence of other polyphenols. For example, the bioavailability of EGCG, the main component of tea catechins, was analyzed when administered as a pure compound, or in an extract. Even though results in rats suggest that the presence of other compounds interferes with bioavailability [50], in humans, similar levels of EGCG were found when the same doses were administered either as a pure compound or in a green tea catechin mixture [43]. Similarly, absorption of EGCG through a monolayer of CACO-2 cells was not affected by the form of administration (as a pure molecule, in a tea extract or in combination with various tea polyphenols) [53]. Thus, it seems that CACO-2 could be a good model to investigate the influence of the food matrix on bioavailability.

The influence of food matrix and other components in which the bioactive compound is administered is especially relevant when considering the effects of natural extracts. Several studies highlight the usefulness of using the CACO-2 system to evaluate the effect of extracts, which are often mixtures of different flavonoids, rather than testing the individual pure compounds. Similar to the above-mentioned differences in luteolin, when administered as an extract, the absorption of quercetin, which is transported through a CACO-2 cell monolayer in culture inserts [54] and can be found in a conjugated form in human plasma [55,56], and quercetin conjugates depends on the source (pure vs shallot extract or apple peel extract) [57]. Likewise, transport through a CACO-2 cell monolayer of a pure solution of the flavonoid aspalathin was lower than that of aspalathin from Rooibos extract solutions [58]. The bioavailability of the flavanone hesperetin in a monolayer of CACO-2 cells has also been shown to be modified by the simultaneous presence of other flavonoids [59]. This change in bioavailability could be due to either an inhibition of the hesperitin efflux (transport from basolateral to the apical side) by inhibition of the BRCP transporter or by affecting phase II metabolism to glucuronides and sulfates, or a combination of both mechanisms [59]. Thus the mechanisms by which extracts affect the modification of various compounds within the extracts are not completely understood and are highly dependent on the flavonoid to be transported and the rest of the extract. For some molecules, the extract seems to have no effect. For example, transport of hawthorn flavonoids was similar when administered as a pure compound or as an extract [60]. However, most of the studies show differences between the pure compounds and the extracts, reinforcing the significance of working with natural mixtures. Remarkably, natural extracts might not only contain mixtures of different monomers but also polymers, whose effects could be of importance. In these sense, the flavan-3-ol components of a procyanidin-rich extract derived from grape seeds (GSPE) and their metabolites were analyzed [30,32]. These studies showed that CACO-2 cells are able to sulfate and to a lesser extent glucuronidate the monomeric catechin components of GSPE. In rats, after an acute dose of the same GSPE, catechin and epicatechin metabolites were also found in the plasma [12], but the glucuronidated forms were the most abundant, likely due to the above-mentioned differences between species. Additionally, procyanidin oligomers are also absorbed by the CACO-2 system, since dimers and trimers are found in the basolateral medium. Deprez et al. showed that using CACO-2 cells cultured in inserts, procyanidin dimers and trimers, but not polymers (average polymerization degree of six), are found in the basolateral media. The authors suggested that these compounds are likely absorbed trough the paracellular route [61]. Procyanidin dimers and trimers were also found in the plasma of GSPE-treated rats [12,62]. Thus CACO-2 cells act not only by metabolizing polyphenols but also as a filter to discriminate between the polymers that might actually reach the bloodstream.

The goal of this chapter is to discuss whether the co-culture system resembles the *in vivo* situation. Thus it has to be considered whether the co-culture of CACO-2 cells with other cell types affects their absorption and metabolism. In parallel with the study with GSPE, the effect of co-culturing CACO-2 cells grown on inserts and other cell types, i. e., hepatocytes (HEPG2) and pancreatic beta cells (INS-1), was also analyzed [30,32]. The co-culture system led to no statistically significant differences among the analyzed procyanidins in the apical and basolateral mediums compared to CACO-2 grown alone, indicating that the adaptations found in the co-culture (mentioned above) did not affect the ability of CACO-2 cells to metabolize the extract.

WORKING WITH NATURAL EXTRACTS: ARE THESE FILTERED EXTRACTS BIOACTIVE?

In this chapter it has been examined whether the co-culture of CACO-2 cells and other cell types is viable, whether the cells maintain their functionality, and whether they can serve as a filter to test natural extracts effects on target cells.

Finally, a few studies have evaluated whether this co-culture is useful for bioactivity assays of natural extracts on target cells.

To start with, in the co-culture system of CACO-2 and HEPG2 cells [30], it was tested whether the effects of pure compounds on HEPG2 cells could be reproduced in the co-culture system using EGCG, which was found both in the basolateral side of CACO-2 monolayers and in human plasma [43]. EGCG affects the AMPc stimulation of glucose-6-phosphatase expression on primary hepatocytes [63] and on HEPG2 cells.

These effects were reproduced, although to a much lower degree, by adding a much higher dose (20-fold the effective dose in the HEPG2 monolayer) to the apical face of CACO-2 cells. A two-fold higher concentration showed no effects on HEPG2 cells. These results agree with the observation that EGCG has a very low oral bioavailability in both rats and humans [44].

Concerning whole extracts, the previously described effects of GSPE at decreasing the synthesis and secretion of triglycerides in HEPG2 cells [26] were also reproduced by applying the GSPE extract to the co-culture. GSPE is hypotrigliceridemic *in vivo* [25], and it was confirmed that the GSPE components responsible for the effect on hepatocytes were oligomeric forms that were able to cross the enterocyte barrier [12]. These results are consistent with the results found in other cell lines in which GSPE was fractionated to assess the effects of the fractions; the fractions responsible for the effects of GSPE are oligomeric forms of the compound [64].

Conclusion

The co-culture of CACO-2 cells seeded in a monolayer over other cell types is a feasible system, but it does induce some adaptations that must be considered. Some cell types such as the pancreatic beta-cells INS-1 do not support long-term co-culture, therefore only short time experiments can be carried out. Other cells types like HEPG2 can tolerate long-term co-culture, but with gene expression changes and a modified response to some stimuli, which highlights the importance of using controls that have a known effect on the desired function of study.

Application of bioactive molecules on the apical side of a CACO-2 cell monolayer simulates the human intestinal barrier, and for several bioactive compounds, leads to the metabolites that are found in human plasma. Therefore, it is a more reliable approach than directly applying pure molecules to the target cells. Furthermore, testing the effects of a natural extract on a CACO-2 monolayer is a more suitable approach than using pure compounds because absorption and metabolism might be affected by other components in the extract.

The few experiments carried out with natural extracts and co-culture systems suggest that co-culture does not interfere with CACO-2 cells' ability to act as an intestinal barrier, and reproduce already described effects *in vivo* and *in vitro*, reinforcing the validity of the system as an approach that more closely models *in vivo* situations to test the effect of natural extracts on target cells.

Acknowledgments

The authors are members of the Nutrigenomics Group of the Universitat Rovira i Virgili (http://nutrigenomica.urv.cat). This study was supported by grant number AGL2008-01310 from the Spanish Government.

REFERENCES

[1] Aggett P, Antoine J, Asp N, et al. *PASSCLAIM: consensus on criteria.* Eur. J. Nutr. 2005;44 Suppl 1:i5-30.

[2] Hackman RMF, H., Keen CL. *Flavanols: digestion, absorption and bioactivity.* Phytochem Rev 2008;7:195-208.

[3] Manach C, Williamson G, Morand C, Scalbert A, Rémésy C. *Bioavailability and bioefficacy of polyphenols in humans.* I. Review of 97 bioavailability studies. Am J Clin Nutr 2005;81:230S-242S.

[4] Pinent M, Castell A, Baiges I, et al. *Bioactivity of flavonoids on insulin-secreting cells.* Compr Rev Food Sci F 2008;7:299-308

[5] Kroon P, Clifford M, Crozier A, et al. *How should we assess the effects of exposure to dietary polyphenols in vitro?* Am J Clin Nutr 2004;80:15-21

[6] Baba S, Osakabe N, Natsume M, Terao J. *Absorption and urinary excretion of procyanidin B2 [epicatechin-(4beta-8)-epicatechin] in rats.* Free Radic Biol Med 2002;33:142-148.

[7] Bose M, Lambert J, Ju J, et al. *The major green tea polyphenol, (-)-epigallocatechin-3-gallate, inhibits obesity, metabolic syndrome, and fatty liver disease in high-fat-fed mice.* J Nutr 2008;138:1677-1683.

[8] Chan K, Zhang L, Zuo Z. *Intestinal efflux transport kinetics of green tea catechins in CACO-2 monolayer model.* J Pharm Pharmacol 2007;59: 395-400.

[9] Mackenzie G, Carrasquedo F, Delfino J, et al. *Epicatechin, catechin, and dimeric procyanidins inhibit PMA-induced NF-kappaB activation at multiple steps in Jurkat T cells.* FASEB J 2004;18:167-169.

[10] Vaidyanathan J, Walle T. *Glucuronidation and sulfation of the tea flavonoid (-)-epicatechin by the human and rat enzymes.* Drug Metab Dispos 2002;30:897-903.

[11] García-Ramírez B, Fernandez-Larrea J, Salvadó M, et al. *Tetramethylated dimeric procyanidins are detected in rat plasma and liver early after oral administration of synthetic oligomeric procyanidins.* J Agric Food Chem 2006;54:2543-2551.

[12] Serra A, Macià A, Romero M, et al. *Determination of procyanidins and their metabolites in plasma samples by improved liquid chromatography-tandem mass spectrometry.* J Chromatogr B Analyt Technol Biomed Life Sci 2009;877:1169-1176.

[13] Kris-Etherton P, Hu F, Ros E, Sabaté J. *The role of tree nuts and peanuts in the prevention of coronary heart disease: multiple potential mechanisms.* J Nutr 2008;138:1746S-1751S.

[14] Artursson P, Palm K, Luthman K. *CACO-2 monolayers in experimental and theoretical predictions of drug transport.* Adv Drug Deliv Rev 2001;46:27-43.

[15] Yamashita S, Furubayashi T, Kataoka M, et al. *Optimized conditions for prediction of intestinal drug permeability using CACO-2 cells.* Eur J Pharm Sci 2000;10:195-204.

[16] Hidalgo I, Raub T, Borchardt R. *Characterization of the human colon carcinoma cell line (CACO-2) as a model system for intestinal epithelial permeability.* Gastroenterology 1989;96:736-749.

[17] Artursson P. *Cell cultures as models for drug absorption across the intestinal mucosa.* Crit Rev Ther Drug Carrier Syst 1991;8:305-330.

[18] Sambuy Y, De Angelis I, Ranaldi G, et al. *The CACO-2 cell line as a model of the intestinal barrier: influence of cell and culture-related factors on CACO-2 cell functional characteristics*. Cell Biol Toxicol 2005;21:1-26.

[19] Rubas W, Jezyk N, Grass G. *Comparison of the permeability characteristics of a human colonic epithelial (CACO-2) cell line to colon of rabbit, monkey, and dog intestine and human drug absorption*. Pharm Res 1993;10:113-118.

[20] Tsukura Y, Mori M, Hirotani Y, et al. *Effects of capsaicin on cellular damage and monolayer permeability in human intestinal CACO-2 cells*. Biol Pharm Bull 2007;30:1982-1986.

[21] Suganami T, Nishida J, Ogawa Y. *A paracrine loop between adipocytes and macrophages aggravates inflammatory changes: role of free fatty acids and tumor necrosis factor alpha*. Arterioscler Thromb Vasc Biol 2005;25:2062-2068.

[22] Nollevaux G, Devillé C, El Moualij B, et al. *Development of a serum-free co-culture of human intestinal epithelium cell-lines (CACO-2/HT29-5M21)*. BMC Cell Biol 2006;7:20.

[23] Sakai Y, Arai T, Sakoda A, Suzuki M. *Development of a Simple Double-layered Cell Culture System Using CACO-2 and TIG-1 Cells as a New Cytotoxicity Test*. AATEX 2001;7:47-58.

[24] Choi S, Nishikawa M, Sakoda A, Sakai Y. *Feasibility of a simple double-layered coculture system incorporating metabolic processes of the intestine and liver tissue: application to the analysis of benzo[a]pyrene toxicity*. Toxicol In Vitro 2004;18:393-402.

[25] Del Bas J, Fernández-Larrea J, Blay M, et al. *Grape seed procyanidins improve atherosclerotic risk index and induce liver CYP7A1 and SHP expression in healthy rats*. FASEB J 2005;19:479-481.

[26] Del Bas J, Ricketts M, Vaqué M, et al. *Dietary procyanidins enhance transcriptional activity of bile acid-activated FXR in vitro and reduce triglyceridemia in vivo in a FXR-dependent manner*. Mol Nutr Food Res 2009;53:805-814.

[27] Llópiz N, Puiggròs F, Céspedes E, et al. *Antigenotoxic effect of grape seed procyanidin extract in Fao cells submitted to oxidative stress*. J Agric Food Chem 2004;52:1083-1087.

[28] Montagut G, Bladé C, Blay M, et al. *Effects of a grapeseed procyanidin extract (GSPE) on insulin resistance*. J Nutr Biochem 2009;21:961-967.

[29] Pinent M, Castell A, Baiges I, et al. *Bioactivity of flavonoids on insulin-secreting cells*. Comprehensive Reviews in Food Science and Food Safety 2008;7:299-308.

[30] Castell A, Motilva MJ, Macià A, et al. *Organotypic co-culture system to study plant extract bioactivity on hepatocytes*. In: Food Chem; 2010:775-781.

[31] Markowska M, Oberle R, Juzwin S, et al. *Optimizing CACO-2 cell monolayers to increase throughput in drug intestinal absorption analysis*. J Pharmacol Toxicol Methods 2001;46:51-55.

[32] Castell-Auví A, Cedó L, Pallarés V, et al. *Development of a Coculture System to Evaluate the Bioactivity of Plant Extracts on Pancreatic beta-Cells*. Planta Med 2010;76:1576-1581.

[33] Khoursandi S, Scharlau D, Herter P, et al. *Different modes of sodium-D-glucose cotransporter-mediated D-glucose uptake regulation in CACO-2 cells*. Am J Physiol Cell Physiol 2004;287:C1041-1047.

[34] Walgren R, Lin J, Kinne R, Walle T. *Cellular uptake of dietary flavonoid quercetin 4'-beta-glucoside by sodium-dependent glucose transporter SGLT1.* J Pharmacol Exp Ther 2000;294:837-843.
[35] Kottra G, Daniel H. *Flavonoid glycosides are not transported by the human Na+/glucose transporter when expressed in Xenopus laevis oocytes, but effectively inhibit electrogenic glucose uptake.* J Pharmacol Exp Ther 2007;322:829-835.
[36] Satsu H, Ishimoto Y, Nakano T, et al. *Induction by activated macrophage-like THP-1 cells of apoptotic and necrotic cell death in intestinal epithelial CACO-2 monolayers via tumor necrosis factor-alpha.* Exp Cell Res 2006;312:3909-3919.
[37] Tanoue T, Nishitani Y, Kanazawa K, Hashimoto T, Mizuno M. *In vitro model to estimate gut inflammation using co-cultured CACO-2 and RAW264.7 cells.* Biochem Biophys Res Commun 2008;374:565-569.
[38] Manach C, Hubert J, Llorach R, Scalbert A. *The complex links between dietary phytochemicals and human health deciphered by metabolomics.* Mol Nutr Food Res 2009;53:1303-1315.
[39] Galijatovic A, Otake Y, Walle U, Walle T. *Extensive metabolism of the flavonoid chrysin by human CACO-2 and Hep G2 cells.* Xenobiotica 1999;29:1241-1256.
[40] Galijatovic A, Otake Y, Walle U, Walle T. *Induction of UDP-glucuronosyltransferase UGT1A1 by the flavonoid chrysin in CACO-2 cells-potential role in carcinogen bioinactivation.* Pharm Res 2001;18:374-379.
[41] Galijatovic A, Walle U, Walle T. *Induction of UDP-glucuron osyl transferase by the flavonoids chrysin and quercetin in CACO-2 cells.* Pharm Res 2000;17:21-26.
[42] Gardana C, Simonetti P, Berti C, Pietta P. *Evaluation of propolis polyphenols absorption in humans by liquid chromatography/tandem mass spectrometry.* Rapid Commun Mass Spectrom 2007;21:3849-3854.
[43] Chow H, Cai Y, Alberts D, et al. *Phase I pharmacokinetic study of tea polyphenols following single-dose administration of epigallocatechin gallate and polyphenon E.* Cancer Epidemiol Biomarkers Prev 2001;10:53-58.
[44] Zhang L, Zheng Y, Chow M, Zuo Z. *Investigation of intestinal absorption and disposition of green tea catechins by CACO-2 monolayer model.* Int J Pharm 2004;287:1-12.
[45] Kuhnle G, Spencer J, Schroeter H, et al. *Epicatechin and catechin are O-methylated and glucuronidated in the small intestine.* Biochem Biophys Res Commun 2000;277:507-512.
[46] Brand W, van der Wel P, Rein M, et al. *Metabolism and transport of the citrus flavonoid hesperetin in CACO-2 cell monolayers.* Drug Metab Dispos 2008;36:1794-1802.
[47] Shimoi K, Okada H, Furugori M, et al. *Intestinal absorption of luteolin and luteolin 7-O-beta-glucoside in rats and humans.* FEBS Lett 1998;438:220-224.
[48] Wittemer S, Ploch M, Windeck T, et al. *Bioavailability and pharmacokinetics of caffeoylquinic acids and flavonoids after oral administration of Artichoke leaf extracts in humans.* Phytomedicine 2005;12:28-38.
[49] Mukinda J, Syce J, Fisher D, Meyer M. *Effect of the plant matrix on the uptake of luteolin derivatives-containing Artemisia afra aqueous-extract in CACO-2 cells.* J Ethnopharmacol 2010;130:439-449.

[50] Chen L, Lee M, Li H, Yang C. *Absorption, distribution, elimination of tea polyphenols in rats*. Drug Metab Dispos 1997;25:1045-1050.

[51] Faria A, Pestana D, Azevedo J, et al. *Absorption of anthocyanins through intestinal epithelial cells - Putative involvement of GLUT2*. Mol Nutr Food Res 2009;53:1430-1437.

[52] Chow H, Hakim I, Vining D, et al. *Effects of dosing condition on the oral bioavailability of green tea catechins after single-dose administration of Polyphenon E in healthy individuals*. Clin Cancer Res 2005;11:4627-4633.

[53] Zhang L, Chow M, Zuo Z. *Effect of the co-occurring components from green tea on the intestinal absorption and disposition of green tea polyphenols in CACO-2 monolayer model*. J Pharm Pharmacol 2006;58:37-44.

[54] Walgren R, Walle U, Walle T. *Transport of quercetin and its glucosides across human intestinal epithelial CACO-2 cells*. Biochem Pharmacol 1998;55:1721-1727.

[55] Hollman P, van Trijp J, Buysman M, et al. *Relative bioavailability of the antioxidant flavonoid quercetin from various foods in man*. FEBS Lett 1997;418:152-156.

[56] Jin F, Nieman D, Shanely R, et al. *The variable plasma quercetin response to 12-week quercetin supplementation in humans*. Eur J Clin Nutr 2010;64:692-697.

[57] Boyer J, Brown D, Liu R. *Uptake of quercetin and quercetin 3-glucoside from whole onion and apple peel extracts by CACO-2 cell monolayers*. J Agric Food Chem 2004;52:7172-7179.

[58] Huang M, du Plessis J, du Preez J, Hamman J, Viljoen A. *Transport of aspalathin, a Rooibos tea flavonoid, across the skin and intestinal epithelium*. Phytother Res 2008;22:699-704.

[59] Brand W, Padilla B, van Bladeren P, Williamson G, Rietjens I. *The effect of co-administered flavonoids on the metabolism of hesperetin and the disposition of its metabolites in CACO-2 cell monolayers*. Mol Nutr Food Res 2010;54:851-860.

[60] Zuo Z, Zhang L, Zhou L, Chang Q, Chow M. *Intestinal absorption of hawthorn flavonoids-in vitro, in situ and in vivo correlations*. Life Sci 2006;79:2455-2462.

[61] Deprez S, Mila I, Huneau J, Tome D, Scalbert A. *Transport of proanthocyanidin dimer, trimer, and polymer across monolayers of human intestinal epithelial CACO-2 cells*. Antioxid Redox Signal 2001;3:957-967.

[62] Serra A, Macià A, Romero M, et al. *Bioavailability of procyanidin dimers and trimers and matrix food effects in in vitro and in vivo models*. Br J Nutr 2010;103:944-952.

[63] Collins Q, Liu H, Pi J, et al. *Epigallocatechin-3-gallate (EGCG), a green tea polyphenol, suppresses hepatic gluconeogenesis through 5'-AMP-activated protein kinase*. J Biol Chem 2007;282:30143-30149.

[64] Montagut G, Baiges I, Valls J, et al. *A trimer plus a dimer-gallate reproduce the bioactivity described for an extract of grape seed procyanidins*. Food Chem 2009;116:265-270.

In: CACO-2 Cells and their Uses
Editor: Megan A. Schulz, pp. 61-75

ISBN 978-61209-917-0

Chapter 4

THE USE OF CACO-2 CELLS IN THE STUDY OF INTESTINAL TRANSPORT

Raquel Martín-Venegas and Ruth Ferrer
Departament de Fisiologia, Facultat de Farmàcia, Universitat de Barcelona
Av. Joan XXIII s/n, 08028 Barcelona, Spain

ABSTRACT

CACO-2 cell cultures are widely used *in vitro* models in several scientific fields such as Nutrition, Physiopathology, Pharmacology, and Toxicology. When cultured onto filters, these cells form a highly polarized epithelium that allows discrimination between apical and basolateral compartments. For this reason, CACO-2 cells constitute a suitable experimental strategy for the study of substrate transport across either the transcellular or paracellular pathways of the intestinal epithelium. Under pre-established experimental conditions, these cultures can be used to estimate intestinal absorption (apical-to-basolateral fluxes) and secretion (basolateral-to-apical fluxes). Moreover, they are also highly used for the functional and molecular characterization of the transport systems specific to the apical and basolateral membranes of the enterocyte. Furthermore, dietary regulation of intestinal transport can also be examined by introducing changes in the composition of the apical compartment. Finally, the paracellular pathway, through tight junctions, can be studied using specific substrates of this route. This chapter discusses the value of experimental approaches using CACO-2 cells to address intestinal absorption.

INTRODUCTION

Intestinal transport is a complex transfer process across the epithelium through the paracellular and/or transcellular pathways. Research into absorptive mechanisms is best performed in a model that contains only absorptive cells, without the confounding contribution of mucus, the lamina propria and the muscularis mucosa. Therefore, much attention is currently paid to epithelial cell cultures for the study of substrate transport mechanisms.

The development of human cell culture systems has been a significant breakthrough in the study of intestinal absorption. In this regard, CACO-2 cells are the most popular cellular model used in intestinal transport studies. This cell line was derived from a moderately well-differentiated human colorectal adenocarcinoma that showed enterocytic and colonocytic characteristics. In conventional culture conditions, these cells proliferate until confluence and then differentiate spontaneously into polarized intestinal cells with apical brush border (Figure 1) and tight junctions between them, and they express intestinal hydrolases on the apical membrane (i.e. sucrose-isomaltase, lactase, aminopeptidase N, dipeptidylpeptidase IV).

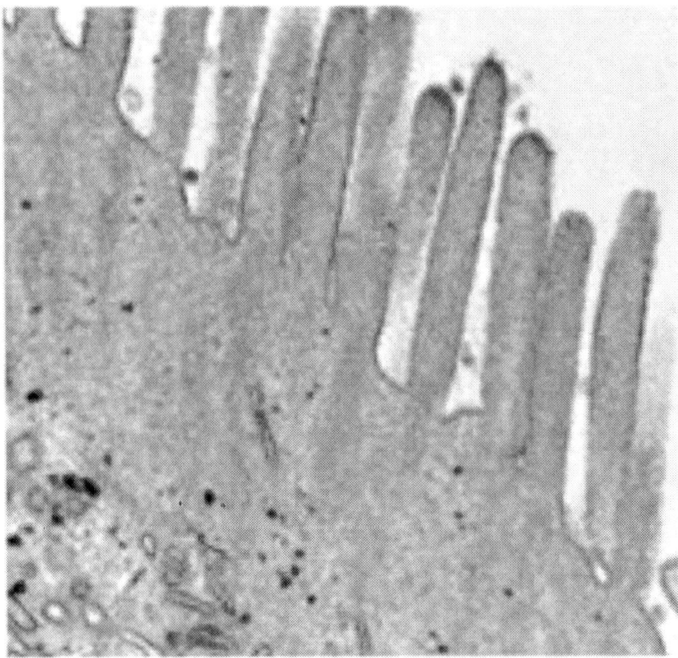

Figure 1. Image obtained by transmission electron microscopy of the brush border membrane of CACO-2 cells after 20 days in culture.

The functional polarity of the cell monolayer is further indicated by the formation, after confluence, of domes, as a result of the presence of unidirectional fluxes of ions and water. Moreover, these cells have polarized membrane receptors for growth factors and several transport activities are present in both membranes. Therefore, despite the colonic origin of these cells, after differentiation they show most of the morphological and functional characteristics of small intestinal absorptive cells [1]. To better reproduce the steric conditions present in the intestine *in vivo*, CACO-2 cells can be cultured on permeable supports that allow separation of an apical and basolateral compartment (Figure 2), which correspond to the intestinal lumen and serosal side, respectively. In these supports, at day 21 CACO-2 cells form a monolayer with the expression of tight junctions, transporters and enzymes.

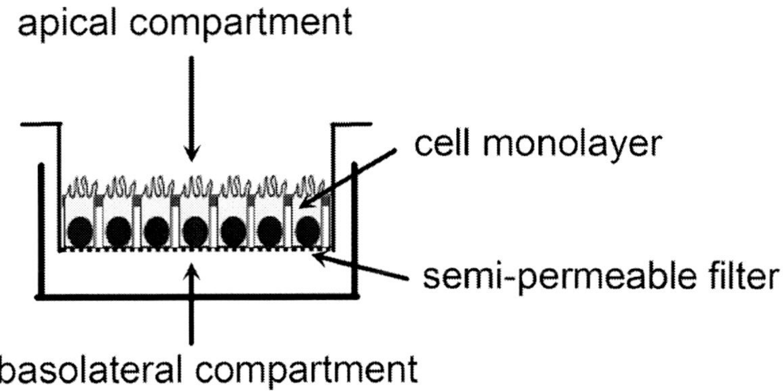

Figure 2. A schematic representation of culture of CACO-2 cells on a microporous filter.

This cell line was first used as a model for studying differentiation in the intestinal epithelium, and later for estimating the relative contributions of paracellular and transcellular passage in substrate absorption. This chapter focuses on the applications of the CACO-2 cell monolayer in intestinal transport.

USES OF CACO-2 CELLS

Intestinal absorption is a complex process that includes two major pathways, namely the paracellular and transcellular routes (Figure 3). The former corresponds to the diffusion of substrates through the intercellular space (i.e. between cells). In contrast, the transcellular route corresponds to the movement of substrates through the cell, first across the apical membrane and then across the basolateral membrane, through both active and passive mechanisms.

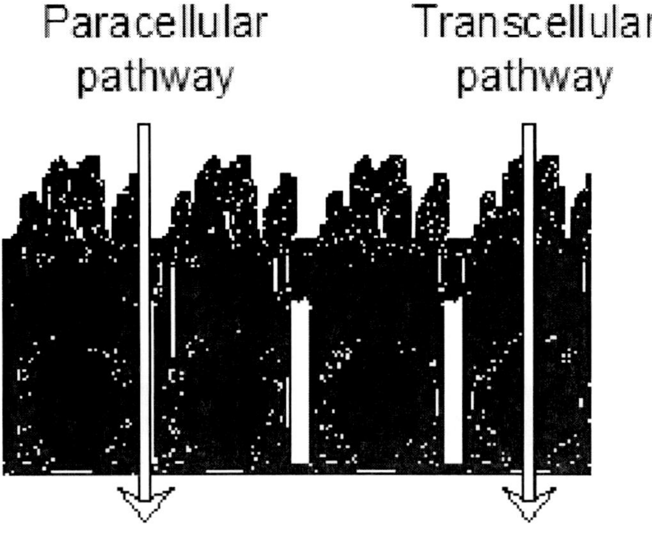

Figure 3. A schematic representation of paracellular and transcellular pathways.

Paracellular Pathway

The structural integrity of the epithelium is maintained by three adhesion systems: tight junctions, adherent junctions, and desmosomes. Of these, tight junctions are the most apical component (Figure 4) and are the rate-limiting step for paracellular permeability. In addition, tight junctions constitute the interface (fence) between apical and basolateral membrane domains [2]. These junctions are multiprotein complexes comprised of transmembrane proteins associated with the cytoskeletal perijunctional ring of actin and myosin and with cytosolic proteins involved in cell signalling and vesicle trafficking [3].

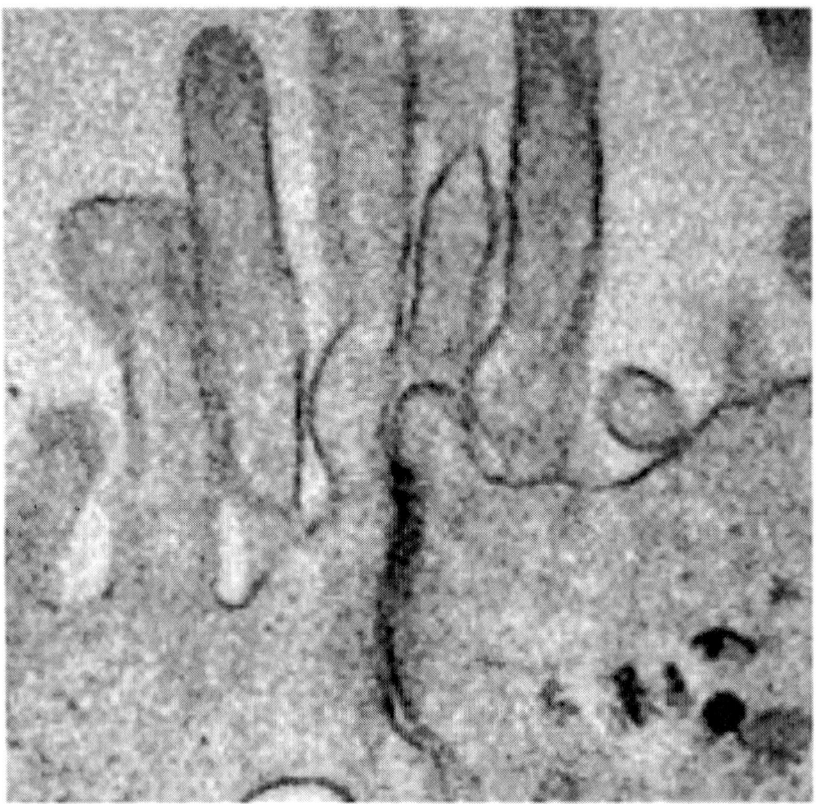

Figure 4. Image obtained by transmission electron microscopy of a tight junction of CACO-2 cells after 20 days in culture.

The paracellular permeability of CACO-2 cultures is routinely checked by measuring transepithelial electrical resistance (TER) and, more reliably, by examining the permeability of paracellular markers such as D-mannitol, inulin, dextran, polyethylene glycol (PEG), and lucifer yellow [4] which are non-metabolizable substrates of diverse molecular weights that cross the intestinal epithelium only through the paracellular pathway. The tightness of the monolayer is indicated by high TER values and low permeability of paracellular markers.

TER determination is a non-destructive method to evaluate membrane potential and resistance of epithelial cells to ion fluxes through the paracellular pathway. TER is measured mostly as described by Hidalgo *et al.* [5], using a voltohmmeter equipped with a pair of electrodes that differ in size to facilitate their placement in cell culture wells.

CACO-2 cell cultures were first used to study paracellular permeability, with the objective to improve the absorption of hydrophilic drugs that cannot be transported across the transcellular pathway. In these first experiments, epithelial barrier function was disrupted by a Ca^{2+} switch assay [6]. These first results paved the way for numerous studies, many performed in CACO-2 cell cultures, which have significantly improved knowledge about tight junction structure and permeability. The role of Ca^{2+} was further addressed [7] along with the identification of the intracellular signaling pathways, such as protein kinase C (PKC), myosin light chain kinase and Rho kinase, all of which involved in the regulation of epithelial barrier function [8-13]. In addition, the structural and functional characterization of tight junction and cytoskeletal proteins have been reported in CACO-2 cell cultures [14-16].

In the first studies concerning the regulation of paracellular permeability it was already suspected that the impairment of epithelial barrier function might contribute to the etiology of different gastrointestinal pathologies. Once again, CACO-2 cells were crucial to confirm this hypothesis. In this regard, much attention has been devoted to the effect of several cytokines on the modulation of paracellular permeability. The pro-inflammatory cytokines, tumor necrosis factor-α, interferon-γ and interleukin-1β (TNF-α, IFN-γ and IL-1β), produced by intestinal wall lymphocytes and mononuclear cells, were found to be responsible for the increase in paracellular permeability that causes the onset or perpetuation of inflammation [17-19]. In addition, the mechanisms responsible for the effect of these cytokines and also preventive therapeutic strategies have been studied in CACO-2 cell cultures [20-28]. Although the disruption of the epithelial barrier function observed in inflammatory bowel disease (IBD) patients has been traditionally attributed to cytokines, other inflammatory mediators such as PGE_2, which is increased in the mucosa of IBD patients [29], has also been involved in the regulation of paracellular permeability in CACO-2 cells [30-32].

Under increased paracellular permeability, toxins, allergenic compounds and bacteria may enter systemic tissues, resulting in inflammatory or immunologic responses [33]. In this regard, the CACO-2 cell system is an efficient model to study the epithelial barrier function of several infectious microorganisms that are also involved in gastrointestinal disorders, such as rotavirus, *Escherichia coli*, *Vibrio parahaemolyticus* and *Salmonella* [34-38]. Moreover, the role of the toll-like receptor 2 pathway, which plays a key role in microbial recognition and immune modulation in the regulation of tight junction permeability, has also been described in this experimental model [39]. Similarly, the protective effect of probiotics, such as *Lactobacillus rhamnosus* and *Lactobacillus casei,* against pro-inflammatory cytokines has also been demonstrated in CACO-2 cells [40,41].

In addition, several studies on the regulation of paracellular permeability have focused on dietary components. Thus, the effects of ethanol [42], fatty acids [43-48] and flavonoids [49], among others, have also been examined in this cell line. Finally, some of the above described effects on epithelial barrier function have been attributed to oxidative stress. The production of reactive oxygen and nitrogen species (ROS and RNS) has gained relevance in recent years in the context of the regulation of tight junction permeability [50-55].

Transcellular Pathway

Enterocytes from the small intestine express several carrier proteins on apical and basolateral membranes that allow transepithelial passage of nutritionally relevant molecules from the intestinal lumen to circulating blood. The CACO-2 cell monolayer is equipped with most of the known intestinal transporters in a rank order or pattern that is similar to that of the small intestine [56,57]. In this regard, specific intestinal carriers for sugars, amino acids, di- and tri-peptides, fatty acids, vitamins, bile acids, heavy metals and nucleotides have been identified and functionally characterized in CACO-2 cells and more recently, several of them have also been ascribed to specific gene products. Therefore, because most intestinal transporters are expressed in the CACO-2 monolayer, this cell line provides a useful model to study intestinal transepithelial transport.

The first study on monosaccharide transport in CACO-2 cell cultures was published in 1987 by Blais *et al.* [58]. These authors functionally characterized active Na^+-dependent monosaccharide transport across the apical membrane during cell differentiation. Since then, several studies have extended our knowledge of the apical and basolateral mechanisms involved in intestinal monosaccharide absorption. Years after that first pioneering study, the distribution of the Na^+-dependent glucose transport system (SGLT1) and the facilitated carriers (GLUT family) in these two membranes were described [59,60]. Furthermore, also in CACO-2 cells, it has been reported that the largest amount of SGLT1 resides in intracellular compartments, which represent an intracellular reserve pool [61] and that D-glucose transport is pharmacologically regulated by a shift of SGLT1 protein between the plasma membrane and the endosomal pool [62]. Another milestone achieved in these cells for intestinal monosaccharide transport is that the increase observed in tight junction permeability induced by SGLT1 activity is mediated by myosin light chain phosphorylation and thus, by the contraction of the perijunctional actomyosin ring [63].

Short-chain fatty acids are produced in abundance in the colon. Given the origin of CACO-2 cell cultures, these cells have been extensively used to study the transport of short-chain fatty acids and other monocarboxylic acids by the monocarboxylate transport family MCT [64-69], which has also been implicated in the absorption of non-steroidal anti-inflammatory drugs (NSAIDs) and the hydroxy analog of L-methionine [70,71].

Regarding amino acid transport, the CACO-2 cell system was also used to characterize the transport systems involved [72-84]. In 1995, Thwaites *et al.* [85] described for the first time a H^+-dependent transport system, system PAT1, in the apical membrane of CACO-2 cells. Since then, most studies undertaken to characterize this transport mechanism and the contribution of this system to the absorption of pharmacologically active compounds have been performed in this experimental model [86-94].

Peptide transporters are expressed mainly in the small intestine and kidney (PepT1 and PepT2, respectively). The endogenous substrates for these transporters are small peptides consisting of two or three amino acids. The main functional characteristics of system PepT1 have been described in cultures of CACO-2 cells [95-103]. In addition, the contribution of these cells has been highly relevant in the discovery of the capacity of PepT1 to transport pharmacologically active compounds, such as β-lactam antibiotics [104-108], antivirals [109-111], L-dopa [112], antipsychotic agents [113] and bacterial derived pro-inflammatory n-formyl peptides [114] among others.

CONCLUSION

Although the translation of *in vitro* CACO-2 data to *in vivo* conditions requires further research, CACO-2 cell cultures have become a useful tool to evaluate intestinal transport. Despite the availability of genetically engineered cell lines, the CACO-2 cell monolayer continues to be the model of choice for the assessment of intestinal transport. This choice is attributed to the characteristics of the cell line; the system is enterocyte-like, well characterized, and easily maintained and manipulated. These characteristics account for this system being one of the most widely used in fields such as Nutrition, Physiopathology, Pharmacology, and Toxicology.

REFERENCES

[1] Grasset, E., Pinto, M., Dussaulx, E., Zweibaum, A. & Desjeux, J.F. (1984). *Epithelial properties of human colonic carcinoma cell line CACO-2: electrical parameters.* Am. J. Physiol., 247, C260-C267.

[2] Mitic, L.L. & Anderson, J.M. (1998). *Molecular architecture of tight junctions.* Annu. Rev. Physiol., 60, 121-142.

[3] Shin, K., Fogg, V.C. & Margolis, B. (2006). *Tight junctions and cell polarity.* Annu. Rev. Cell Dev. Biol., 22, 207-235.

[4] Mukherjee, T., Squillantea, E., Gillespieb, M. & Shao, J. (2004). *Transepithelial electrical resistance is not a reliable measurement of the CACO-2 monolayer integrity in Transwell.* Drug Deliv., 11, 11-18.

[5] Hidalgo, I.J., Raub, T.J. & Borchardt, R.T. (1989). *Characterization of the human colon carcinoma cell line (CACO-2) as a model system for intestinal epithelial permeability.* Gastroenterology, 96, 736-749.

[6] Artursson, P. & Magnusson, C. (1990). *Epithelial transport of drugs in cell culture. II: Effect of extracellular calcium concentration on the paracellular transport of drugs of different lipophilicities across monolayers of intestinal epithelial (CACO-2) cells.* J. Pharm. Sci., 79, 595-600.

[7] Rutten, M.J., Cogburn, J.N., Schasteen, C.S. & Solomon, T. (1991). *Physiological and cytotoxic effects of Ca^{2+} ionophores on CACO-2 paracellular permeability: relationship of $^{45}Ca^{2+}$ efflux to ^{51}Cr release.* Pharmacology, 42, 156-168.

[8] Stenson, W.F., Easom, R.A., Riehl, T.E. & Turk, J. (1993). *Regulation of paracellular permeability in CACO-2 cell monolayers by protein kinase C.* Am. J. Physiol., 265, G955-G962.

[9] Ma, T.Y., Tran, D., Hoa, N., Nguyen, D., Merryfield, M. & Tarnawski, A. (2000). *Mechanism of extracellular calcium regulation of intestinal epithelial tight junction permeability: role of cytoskeletal involvement.* Microsc. Res. Tech., 51, 156-168.

[10] Walsh, S.V., Hopkins, A.M., Chen, J., Narumiya, S., Parkos, C.A. & Nusrat, A. (2001). *Rho kinase regulates tight junction function and is necessary for tight junction assembly in polarized intestinal epithelia.* Gastroenterology, 121, 566-579.

[11] Zolotarevsky, Y., Hecht, G., Koutsouris, A., Gonzalez, D.E., Quan, C., Tom, J., Mrsny, R.J. & Turner, J.R. (2002). *A membrane-permeant peptide that inhibits MLC kinase*

restores barrier function in in vitro models of intestinal disease. Gastroenterology, 123, 163-172.

[12] Clayburgh, D.R., Rosen, S., Witkowski, E.D., Wang, F., Blair, S., Dudek, S., Garcia, J.G., Alverdy, J.C. & Turner, J.R. (2004). *A differentiation-dependent splice variant of myosin light chain kinase, MLCK1, regulates epithelial tight junction permeability*. J. Biol. Chem., 31, 55506-55513.

[13] Suzuki, T., Elias, B.C., Seth, A., Shen, L., Turner, J.R., Giorgianni, F., Desiderio, D., Guntaka, R. & Rao, R. (2009). *PKC eta regulates occludin phosphorylation and epithelial tight junction integrity*. Proc. Natl. Acad. Sci. USA, 106, 61-66.

[14] Collares-Buzato, C.B., McEwan, G.T., Jepson, M.A., Simmons, N.L. & Hirst, B.H. (1994). *Paracellular barrier and junctional protein distribution depend on basolateral extracellular* Ca^{2+} *in cultured epithelia*. Biochim. Biophys. Acta, 1222, 147-158.

[15] Ma, T.Y., Hollander, D., Tran, L.T., Nguyen, D., Hoa, N. & Bhalla, D. (1995). *Cytoskeletal regulation of CACO-2 intestinal monolayer paracellular permeability*. J. Cell. Physiol., 164, 533-545.

[16] Turner, J.R., Angle, J.M., Black, E.D., Joyal, J.L., Sacks, D.B. & Madara, J.L. (1999). *PKC-dependent regulation of transepithelial resistance: roles of MLC and MLC kinase*. Am. J. Physiol., 277, C554-C562.

[17] Andus, T., Gross, V., Casar, I., Krumm, D., Hosp, J., David, M. & Schölmerich, J. (1991). *Activation of moncytes during inflammatory bowel diease*. Pathobiology, 59, 166-170.

[18] Lu, J., Philpott, D.J., Saunders, P.R., Perdue, M.H., Yang, P.C. & McKay, D.M. (1998). *Epithelial ion transport and barrier abnormalities evoked by superantigen-activated immune cells are inhibited by interleukin-10 bot not interleukin-4*. J. Pharmacol. Exp. Ther., 287, 128-136.

[19] Al-Sadi, R.M. & Ma, T.Y. (2007). *IL-1beta causes an increase in intestinal epithelial tight junction permeability*. J. Immunol., 178, 4641-4649.

[20] Di Paolo, M.C., Merrett, M.N., Crotty, B. & Jewell, D.P. (1996). *5-Aminosalicylic acid inhibits the impaired epithelial barrier function induced by gamma interferon*. Gut, 38, 115-119.

[21] Marano, C.W., Lewis, S.A., Garulacan, L.A., Soler, A.P. & Mullin, J.M. (1998). *Tumor necrosis factor-alpha increases sodium and chloride conductance across the tight junction of CACO-2 BBE, a human intestinal epithelial cell line*. J. Membr. Biol., 161, 263-274.

[22] Rao, R., Baker, R.D. & Baker, S.S. (1999). *Inhibition of oxidant-induced barrier disruption and protein tyrosine phosphorylation in CACO-2 cell monolayers by epidermal growth factor*. Biochem. Pharmacol., 57, 685-695.

[23] Han, X., Fink, M.P. & Delude, R.L. (2003). *Proinflammatory cytokines cause NO•-dependent and -independent changes in expression and localization of tight junction proteins in intestinal epithelial cells*. Shock, 19, 229-237.

[24] Wang, F., Graham, W.V., Wang, Y., Witkowski, E.D., Schwarz, B.T. & Turner, J.R. (2005). *Interferon-gamma and tumor necrosis factor-alpha synergize to induce intestinal epithelial barrier dysfunction by up-regulating myosin light chain kinase expression*. Am. J. Pathol., 166, 409-419.

[25] Ye, D., Ma, I. & Ma, T.Y. (2006). *Molecular mechanism of tumor necrosis factor-alpha modulation of intestinal epithelial tight junction barrier.* Am. J. Physiol. Gastrointest. Liver Physiol., 290, G496-G504.

[26] Schwarz, B.T., Wang, F., Shen, L., Clayburgh, D.R., Su, L., Wang, Y., Fu, Y.X. & Turner, J.R. (2007). *LIGHT signals directly to intestinal epithelia to cause barrier dysfunction via cytoskeletal and endocytic mechanisms.* Gastroenterology, 132, 2383-2394.

[27] Wang, Q., Guo, X.L., Wells-Byrum, D., Noel, G., Pritts, T.A. & Ogle, C.K. (2008). *Cytokine-induced epithelial permeability changes are regulated by the activation of the p38 mitogen-activated protein kinase pathway in cultured CACO-2 cells.* Shock, 29, 531-537.

[28] Al-Sadi, R., Ye, D., Said, H.M. & Ma, T.Y. (2010). *IL-1beta-induced increase in intestinal epithelial tight junction permeability is mediated by MEKK-1 activation of canonical NF-kappaB pathway.* Am. J. Pathol., 177, 2310-2322.

[29] Hommes, D.W., Meenan, J., de Haas, M., ten Kate, F.J., von dem Borne, A.E., Tytgat, G.N. & van Deventer, S.J. (1996). *Soluble Fc gamma receptor III (CD 16) and eicosanoid concentrations in gut lavage fluid from patients with inflammatory bowel disease: reflection of mucosal inflammation.* Gut, 38, 564-567.

[30] Martín-Venegas, R., Roig-Pérez, S., Ferrer, R. & Moreno, J.J. (2006). *Arachidonic acid cascade and epithelial barrier function during CACO-2 cell differentiation.* J. Lipid Res., 47, 1416-1423.

[31] Rodríguez-Lagunas, M.J., Martín-Venegas, R., Moreno, J.J. & Ferrer, R. (2010). *PGE_2 promotes Ca^{2+}-mediated epithelial barrier disruption through EP1 and EP4 receptors in CACO-2 cell monolayers.* Am. J. Physiol. Cell Physiol., 299, C324-C334.

[32] Tanaka, M.N., Diaz, B.L., de Souza, W. & Morgado-Diaz, J.A. (2008). *Prostaglandin E2-EP1 and EP2 receptor signaling promotes apical junctional complex disassembly of CACO-2 human colorectal cancer cells.* BMC Cell Biol., 9, 63.

[33] Deitch, E.A. (1993). *Nutrition and the gut mucosal barrier.* Curr. Opin. Gen. Surg., 1993, 85-91.

[34] Dickman, K.G., Hempson, S.J., Anderson, J., Lippe, S., Zhao, L., Burakoff, R. & Shaw, R.D.. (2000). *Rotavirus alters paracellular permeability and energy metabolism in CACO-2 cells.* Am. J. Physiol. Gastrointest. Liver Physiol., 279, G757-G766.

[35] Obert, G., Peiffer, I. & Servin, A.L. (2000). *Rotavirus-induced structural and functional alterations in tight junctions of polarized intestinal CACO-2 cell monolayers.* J. Virol., 74, 4645-4651.

[36] Peiffer, I., Blanc-Potard, A.B., Bernet-Camard, M.F., Guignot, J., Barbat, A. & Servin, A.L. (2000). *Afa/Dr diffusely adhering Escherichia coli C1845 infection promotes selective injuries in the junctional domain of polarized human intestinal CACO-2/TC7 cells.* Infect. Immun., 68, 3431-3442.

[37] Solano, C., Sesma, B., Alvarez, M., Urdaneta, E., Garcia-Ros, D., Calvo, A. & Gamazo, C. (2001). *Virulent strains of Salmonella enteritidis disrupt the epithelial barrier of CACO-2 and HEp-2 cells.* Arch. Microbiol., 175, 46-51.

[38] Lynch, T., Livingstone, S., Buenaventura, E., Lutter, E., Fedwick, J., Buret, A.G., Graham, D. & DeVinney, R. (2005). *Vibrio parahaemolyticus disruption of epithelial cell tight junctions occurs independently of toxin production.* Infect. Immun., 73, 1275-1283.

[39] Cario, E., Gerken, G. & Podolsky, D.K. (2004). *Toll-like receptor 2 enhances ZO-1-associated intestinal epithelial barrier integrity via protein kinase C.* Gastroenterology, 127, 224-238.

[40] Miyauchi, E., Morita, H. & Tanabe, S. (2009). *Lactobacillus rhamnosus alleviates intestinal barrier dysfunction in part by increasing expression of zonula occludens-1 and myosin light-chain kinase in vivo.* J. Dairy Sci., 92, 2400-2408.

[41] Eun, C.S., Kim, Y.S., Han, D.S., Choi, J.H., Lee, A.R. & Park, Y.K. (2011). *Lactobacillus casei prevents impaired barrier function in intestinal epithelial cells.* APMIS, 119, 49-56.

[42] Ma, T.Y., Nguyen, D., Bui, V., Nguyen, H. & Hoa, N. (1999). *Ethanol modulation of intestinal epithelial tight junction barrier.* Am. J. Physiol. 276, G965-G974.

[43] Duizer, E., van der Wulp, C., Versantvoort, C.H. & Groten, J.P. (1998). *Absorption enhancement, structural changes in tight junctions and cytotoxicity caused by palmitoyl carnitine in CACO-2 and IEC-18 cells.* J. Pharmacol. Exp. Ther., 287, 395-402.

[44] Lindmark, T., Kimura, Y. & Artursson, P. (1998). *Absorption enhancement through intracellular regulation of tight junction permeability by medium chain fatty acids in CACO-2 cells.* J. Pharmacol. Exp. Ther., 284, 362-369.

[45] Roche, H.M., Terres, A.M., Black, I.B., Gibney, M.J. & Kelleher, D. (2001). *Fatty acids and epithelial permeability: effect of conjugated linoleic acid in CACO-2 cells.* Gut, 48, 797-802.

[46] Usami, M., Muraki, K., Iwamoto, M., Ohata, A., Matsushita, E. & Miki, A. (2001). *Effect of eicosapentanoic acid (EPA) on tight junction permeability in intestinal monolayer cells.* Clin. Nutr., 20, 351-359.

[47] Roig-Pérez, S., Guardiola, F., Moretó, M. & Ferrer, R. (2004). *Lipid peroxidation induced by DHA enrichment modifies paracellular permeability in CACO-2 cells: protective role of taurine.* J. Lipid Res., 45, 1418-1428.

[48] Roig-Pérez, S., Cortadellas, N., Moretó, M. & Ferrer, R. (2010). *Intracellular mechanisms involved in docosahexaenoic acid-induced increases in tight junction permeability in CACO-2 cell monolayers.* J. Nutr., 140, 1557-1563.

[49] Amasheh, M., Schlichter, S., Amasheh, S., Mankertz, J., Zeitz, M., Fromm, M. & Schulzke, J.D. (2008). *Quercetin enhances epithelial barrier function and increases claudin-4 expression in CACO-2 cells.* J. Nutr., 138, 1067-1073.

[50] Rao, R.K., Baker, R.D., Baker, S.S., Gupta, A. & Holycross, M. (1997). *Oxidant-induced disruption of intestinal epithelial barrier function: role of protein tyrosine phosphorylation.* Am. J. Physiol., 273, G812-G823.

[51] Rao, R., Baker, R.D. & Baker, S.S. (1999). *Inhibition of oxidant-induced barrier disruption and protein tyrosine phosphorylation in CACO-2 cell monolayers by epidermal growth factor.* Biochem. Pharmacol., 15, 685-695.

[52] Rao, R.K., Li, L., Baker, R.D., Baker, S.S. & Gupta, A. (2000). *Glutathione oxidation and PTPase inhibition by hydrogen peroxide in CACO-2 cell monolayer.* Am. J. Physiol. Gastrointest. Liver Physiol., 279, G332-G340.

[53] Banan, A., Fitzpatrick, L., Zhang, Y. & Keshavarzian, A. (2001). *OPC-compounds prevent oxidant-induced carbonylation and depolymerization of the F-actin cytoskeleton and intestinal barrier hyperpermeability.* Free Radic. Biol. Med., 30, 287-298.

[54] Basuroy, S., Seth, A., Elias, B., Naren, A.P. & Rao, R. (2006). *MAPK interacts with occludin and mediates EGF-induced prevention of tight junction disruption by hydrogen peroxide*. Biochem. J., 393, 69-77.

[55] Musch, M.W., Walsh-Reitz, M.M. & Chang, E.B. (2006). *Roles of ZO-1, occludin, and actin in oxidant-induced barrier disruption.* Am. J. Physiol. Gastrointest. Liver Physiol., 290, G222-G231.

[56] Englund, G., Rorsman, F., Rönnblom, A., Karlbom, U., Lazorova, L., Gråsjö, J., Kindmark, A. & Artursson, P. (2006). *Regional levels of drug transporters along the human intestinal tract: co-expression of ABC and SLC transporters and comparison with CACO-2 cells.* Eur. J. Pharm. Sci., 29, 269-277.

[57] Hilgendorf, C., Ahlin, G., Seithel, A., Artursson, P., Ungell, A.L. & Karlsson, J. (2007). *Expression of thirty-six drug transporter genes in human intestine, liver, kidney, and organotypic cell lines.* Drug Metab. Dispos., 35, 1333-1340.

[58] Blais, A., Bissonnette, P. & Berteloot, A. (1987). *Common characteristics for Na^+-dependent sugar transport in CACO-2 cells and human fetal colon.* J. Membr. Biol., 99, 113-125.

[59] Harris, D.S., Slot, J.W., Geuze, H.J. & James, D.E. (1992). *Polarized distribution of glucose transporter isoforms in CACO-2 cells.* Proc. Natl. Acad. Sci. USA, 89, 7556-7560.

[60] Mahraoui, L., Rousset, M., Dussaulx, E., Darmoul, D., Zweibaum, A. & Brot-Laroche, E. (1992). *Expression and localization of GLUT-5 in CACO-2 cells, human small intestine, and colon.* Am. J. Physiol., 263, G312-G318.

[61] Kipp, H., Khoursandi, S., Scharlau, D. & Kinne, R.K. (2003). *More than apical: Distribution of SGLT1 in CACO-2 cells.* Am. J. Physiol. Cell Physiol., 285, C737-C749.

[62] Khoursandi, S., Scharlau, D., Herter, P., Kuhnen, C., Martin, D., Kinne, R.K. & Kipp, H. (2004). *Different modes of sodium-D-glucose cotransporter-mediated D-glucose uptake regulation in CACO-2 cells.* Am. J. Physiol. Cell Physiol., 287, C1041-C1047.

[63] Turner, J.R., Rill, B.K., Carlson, S.L., Carnes, D., Kerner, R., Mrsny, R.J. & Madara, J.L. (1997). *Physiological regulation of epithelial tight junctions is associated with myosin light-chain phosphorylation.* Am. J. Physiol., 273, C1378-C1385.

[64] Tamai, I., Takanaga, H., Maeda, H., Sai, Y., Ogihara, T., Higashida, H. & Tsuji, A. (1995). *Participation of a proton-cotransporter, MCT1, in the intestinal transport of monocarboxylic acids. Biochem.* Biophys. Res. Commun., 214, 482-489.

[65] Hadjiagapiou, C., Schmidt, L., Dudeja, P.K., Layden, T.J. & Ramaswamy, K. (2000). *Mechanism(s) of butyrate transport in CACO-2 cells: role of monocarboxylate transporter 1.* Am. J. Physiol. Gastrointest. Liver Physiol., 279, G775-G780.

[66] Stein, J., Zores, M. & Schröder, O. (2000). *Short-chain fatty acid (SCFA) uptake into CACO-2 cells by a pH-dependent and carrier mediated transport mechanism.* Eur. J. Nutr., 39, 121-125.

[67] Borthakur, A., Saksena, S., Gill, R.K., Alrefai, W.A., Ramaswamy, K. & Dudeja, P.K. (2008). *Regulation of monocarboxylate transporter 1 (MCT1) promoter by butyrate in human intestinal epithelial cells: involvement of NF-kappaB pathway.* J. Cell. Biochem., 103, 1452-1463.

[68] Saksena, S., Dwivedi, A., Gill, R.K., Singla, A., Alrefai, W.A., Malakooti, J., Ramaswamy, K. & Dudeja, P.K. (2009). *PKC-dependent stimulation of the human*

MCT1 promoter involves transcription factor AP2. Am. J. Physiol. Gastrointest. Liver Physiol., 296, G275-G283.

[69] Saksena, S., Theegala, S., Bansal, N., Gill, R.K., Tyagi, S., Alrefai, W.A., Ramaswamy, K. & Dudeja, P.K. (2009). *Mechanisms underlying modulation of monocarboxylate transporter 1 (MCT1) by somatostatin in human intestinal epithelial cells.* Am. J. Physiol. Gastrointest. Liver Physiol., 297, G878-G885.

[70] Choi, J.S., Jin, M.J. & Han, H.K. (2005). *Role of monocarboxylic acid transporters in the cellular uptake of NSAIDs.* J. Pharm. Pharmacol., 57, 1185-1189.

[71] Martín-Venegas, R., Rodríguez-Lagunas, M.J., Geraert, P.A. & Ferrer, R. (2007). *Monocarboxylate transporter 1 mediates DL-2-Hydroxy-(4-methylthio)butanoic acid transport across the apical membrane of CACO-2 cell monolayers.* J. Nutr., 137, 49-54.

[72] Chen, J., Zhu, Y. & Hu, M. (1994). *Mechanisms and kinetics of uptake and efflux of L-methionine in an intestinal epithelial model (CACO-2).* J. Nutr., 124, 1907-1916.

[73] Zheng, L., Chen, J., Zhu, Y., Yang, H., Elmquist, W. & Hu, M. (1994). *Comparison of the transport characteristics of D- and L-methionine in a human intestinal epithelial model (CACO-2) and in a perfused rat intestinal model.* Pharm. Res., 11, 1771-1776.

[74] Pan, M. & Stevens, B.R. (1995). *Differentiation- and protein kinase C-dependent regulation of alanine transport via system B.* J. Biol. Chem., 24, 3582-3587.

[75] Ferruzza, S., Ranaldi, G., Di Girolamo, M. & Sambuy, Y. (1995). *The transport of lysine across monolayers of human cultured intestinal cells (CACO-2) depends on Na^+-dependent and Na^+-independent mechanisms on different plasma membrane domains.* J. Nutr., 125, 2577-2585.

[76] Thwaites, D.T., Markovich, D., Murer, H. & Simmons, N.L. (1996). *Na^+-independent lysine transport in human intestinal CACO-2 cells.* J. Membr. Biol., 151, 215-224.

[77] Kekuda, R., Torres-Zamorano, V., Fei, Y.J., Prasad, P.D., Li, H.W., Mader, L.D., Leibach, F.H. & Ganapathy, V. (1997). *Molecular and functional characterization of intestinal Na^+-dependent neutral amino acid transporter B^0.* Am. J. Physiol., 272, G1463-G1472.

[78] Satsu, H., Watanabe, H., Arai, S. & Shimizu, M. (1998). *System $B^{0,+}$-mediated regulation of lysine transport in CACO-2 human intestinal cells.* Amino Acids, 14, 379-384.

[79] Berger, V., Larondelle, Y., Trouet, A. & Schneider, Y.J. (2000). *Transport mechanisms of the large neutral amino acid L-phenylalanine in the human intestinal epithelial CACO-2 cell line.* J. Nutr., 130, 2780-2788.

[80] Berger, V., De Bremaeker, N., Larondelle, Y., Trouet, A. & Schneider, Y.J. (2000). *Transport mechanisms of the imino acid L-proline in the human intestinal epithelial CACO-2 cell line.* J. Nutr., 130, 2772-2779.

[81] Costa, C., Huneau, J. & Tomé, D. (2000). *Characteristics of L-glutamine transport during CACO-2 cell differentiation.* Biochim. Biophys. Acta, 1509, 95-102.

[82] Christie, G.R., Ford, D., Howard, A., Clark, M.A. & Hirst, B.H. (2001). *Glycine supply to human enterocytes mediated by high-affinity basolateral GLYT1.* Gastroenterology, 120, 439-448.

[83] Fraga, S., Pinho, M.J. & Soares-da-Silva, P. (2005). *Expression of LAT1 and LAT2 amino acid transporters in human and rat intestinal epithelial cells.* Amino Acids, 29, 229-233.

[84] Martín-Venegas, R., Rodríguez-Lagunas, M.J., Mercier, Y., Geraert, P.A. & Ferrer, R. (2009). *Effect of pH on L- and D-methionine uptake across the apical membrane of CACO-2 cells*. Am. J. Physiol. Cell Physiol., 296, C632-C638.

[85] Thwaites, D.T., McEwan, G.T. & Simmons, N.L. (1995). *The role of the proton electrochemical gradient in the transepithelial absorption of amino acids by human intestinal CACO-2 cell monolayers*. J. Membr. Biol., 145, 245-256.

[86] Boll, M., Foltz, M., Anderson, C.M., Oechsler, C., Kottra, G., Thwaites, D.T. & Daniel, H. (2003). *Substrate recognition by the mammalian proton-dependent amino acid transporter PAT1*. Mol. Membr. Biol., 20, 261-269.

[87] Chen, Z., Fei, Y.J., Anderson, C.M., Wake, K.A., Miyauchi, S., Huang, W., Thwaites, D.T. & Ganapathy, V. (2003). *Structure, function and immunolocalization of a proton-coupled amino acid transporter (hPAT1) in the human intestinal cell line CACO-2*. J. Physiol., 546, 349-361.

[88] Anderson, C.M., Grenade, D.S., Boll, M., Foltz, M., Wake, K.A., Kennedy, D.J., Munck, L.K., Miyauchi, S., Taylor, P.M., Campbell, F.C., Munck, B.G., Daniel, H., Ganapathy, V. & Thwaites, D.T. (2004). *H^+/amino acid transporter 1 (PAT1) is the imino acid carrier: An intestinal nutrient/drug transporter in human and rat.* Gastroenterology, 127, 1410-1422.

[89] Metzner, L., Neubert, K. & Brandsch, M. (2006). *Substrate specificity of the amino acid transporter PAT1*. Amino Acids, 31, 111-117.

[90] Anderson, C.M., Howard, A., Walters, J.R., Ganapathy, V. & Thwaites, D.T. (2009). *Taurine uptake across the human intestinal brush-border membrane is via two transporters: H^+-coupled PAT1 (SLC36A1) and Na^+- and Cl^--dependent TauT (SLC6A6)*. J. Physiol., 587, 731-744.

[91] Larsen, M., Holm, R., Jensen, K.G., Brodin, B. & Nielsen, C.U. (2009). *Intestinal gaboxadol absorption via PAT1 (SLC36A1): modified absorption in vivo following co-administration of L-tryptophan*. Br. J. Pharmacol., 157, 1380-1389.

[92] Metzner, L., Dorn, M., Markwardt, F. & Brandsch, M. (2009). *The orally active antihyperglycemic drug beta-guanidinopropionic acid is transported by the human proton-coupled amino acid transporter hPAT1*. Mol. Pharm., 6, 1006-1011.

[93] Nickel, A., Kottra, G., Schmidt, G., Danier, J., Hofmann, T. & Daniel, H. (2009). *Characteristics of transport of selenoamino acids by epithelial amino acid transporters*. Chem. Biol. Interact., 177, 234-241.

[94] Anderson, C.M., Jevons, M., Thangaraju, M., Edwards, N., Conlon, N.J., Woods, S., Ganapathy, V. & Thwaites, D.T. (2010). *Transport of the photodynamic therapy agent 5-aminolevulinic acid by distinct H^+-coupled nutrient carriers coexpressed in the small intestine*. J. Pharmacol. Exp. Ther., 332, 220-228.

[95] Brandsch, M., Brandsch, C., Ganapathy, M.E., Chew, C.S., Ganapathy, V. & Leibach, F.H. (1997). *Influence of proton and essential histidyl residues on the transport kinetics of the H^+/peptide cotransport systems in intestine (PEPT 1) and kidney (PEPT 2)*. Biochim. Biophys. Acta, 1324, 251-262.

[96] Saito, H., Motohashi, H., Mukai, M. & Inui, K. (1997). *Cloning and characterization of a pH-sensing regulatory factor that modulates transport activity of the human H^+/peptide cotransporter, PEPT1*. Biochem. Biophys. Res. Commun., 237, 577-582.

[97] Börner, V., Fei, Y.J., Hartrodt, B., Ganapathy, V., Leibach, F.H., Neubert, K. & Brandsch, M. (1998). *Transport of amino acid aryl amides by the intestinal H^+/peptide cotransport system, PEPT1.* Eur. J. Biochem., 255, 698-702.

[98] Thamotharan, M., Bawani, S.Z., Zhou, X. & Adibi, S.A. (1999). *Hormonal regulation of oligopeptide transporter pept-1 in a human intestinal cell line.* Am. J. Physiol., 276, C821-C826.

[99] Knütter, I., Theis, S., Hartrodt, B., Born, I., Brandsch, M., Daniel, H. & Neubert, K (2001). *A novel inhibitor of the mammalian peptide transporter PEPT1.* Biochemistry, 40, 4454-4458.

[100] Nielsen, C.U., Amstrup, J., Steffansen, B., Frokjaer, S. & Brodin, B. (2001). *Epidermal growth factor inhibits glycylsarcosine transport and hPepT1 expression in a human intestinal cell line.* Am. J. Physiol. Gastrointest. Liver Physiol., 281, G191-G199.

[101] Kennedy, D.J., Leibach, F.H., Ganapathy, V. & Thwaites, D.T. (2002). *Optimal absorptive transport of the dipeptide glycylsarcosine is dependent on functional Na^+/H^+ exchange activity.* Pflügers Arch., 445, 139-146.

[102] Pieri, M., Christian, H.C., Wilkins, R.J., Boyd, C.A. & Meredith, D. (2010). *The apical (hPepT1) and basolateral peptide transport systems of CACO-2 cells are regulated by AMP-activated protein kinase.* Am. J. Physiol. Gastrointest. Liver Physiol., 299, G136-G143.

[103] Ashida, K., Katsura, T., Motohashi, H., Saito, H. & Inui, K. (2002). *Thyroid hormone regulates the activity and expression of the peptide transporter PEPT1 in CACO-2 cells.* Am. J. Physiol. Gastrointest. Liver Physiol., 282, G617-G623.

[104] Ganapathy, M.E., Brandsch, M., Prasad, P.D., Ganapathy, V. & Leibach, F.H. (1995). *Differential recognition of beta-lactam antibiotics by intestinal and renal peptide transporters, PEPT 1 and PEPT 2.* J. Biol. Chem., 270, 25672-25677.

[105] Wenzel, U., Thwaites, D.T. & Daniel, H. (1995). *Stereoselective uptake of beta-lactam antibiotics by the intestinal peptide transporter.* Br. J. Pharmacol., 116, 3021-3027.

[106] Bretschneider, B., Brandsch, M. & Neubert, R. (1999). *Intestinal transport of beta-lactam antibiotics: analysis of the affinity at the H^+/peptide symporter (PEPT1), the uptake into CACO-2 cell monolayers and the transepithelial flux.* Pharm. Res., 16, 55-61.

[107] Biegel, A., Gebauer, S., Hartrodt, B., Brandsch, M., Neubert, K. & Thondorf, I. (2005). *Three-dimensional quantitative structure-activity relationship analyses of beta-lactam antibiotics and tripeptides as substrates of the mammalian H^+/peptide cotransporter PEPT1.* J. Med. Chem., 48, 4410-4419.

[108] Luckner, P. & Brandsch, M. (2005). *Interaction of 31 beta-lactam antibiotics with the H^+/peptide symporter PEPT2: analysis of affinity constants and comparison with PEPT1.* Eur. J. Pharm. Biopharm., 59, 17-24.

[109] Ganapathy, M.E., Huang, W., Wang, H., Ganapathy, V. & Leibach, F.H. (1998). *Valacyclovir: a substrate for the intestinal and renal peptide transporters PEPT1 and PEPT2.* Biochem. Biophys. Res. Commun., 246, 470-475.

[110] Han, H., de Vrueh, R.L., Rhie, J.K., Covitz, K.M., Smith, P.L., Lee, C.P., Oh, D.M., Sadée, W. & Amidon, G.L. (1998). *5'-Amino acid esters of antiviral nucleosides, acyclovir, and AZT are absorbed by the intestinal PEPT1 peptide transporter.* Pharm. Res., 15, 1154-1159.

[111] Ogihara, T., Kano, T., Wagatsuma, T., Wada, S., Yabuuchi, H., Enomoto, S., Morimoto, K., Shirasaka, Y., Kobayashi, S. & Tamai, I. (2009). *Oseltamivir (tamiflu) is a substrate of peptide transporter 1.* Drug Metab. Dispos., 37, 1676-1681.

[112] Tamai, I., Nakanishi, T., Nakahara, H., Sai, Y., Ganapathy, V., Leibach, F.H. & Tsuji, A. (1998). *Improvement of L-dopa absorption by dipeptidyl derivation, utilizing peptide transporter PepT1.* J. Pharm. Sci., 87, 1542-1546.

[113] Watanabe, K., Sawano, T., Terada, K., Endo, T., Sakata, M. & Sato, J. (2002). *Studies on intestinal absorption of sulpiride (1): carrier-mediated uptake of sulpiride in the human intestinal cell line CACO-2.* Biol. Pharm. Bull., 25, 885-890.

[114] Merlin, D., Si-Tahar, M., Sitaraman, S.V., Eastburn, K., Williams, I., Liu, X., Hediger, M.A. & Madara, J.L. (2001). *Colonic epithelial hPepT1 expression occurs in inflammatory bowel disease: transport of bacterial peptides influences expression of MHC class 1 molecules.* Gastroenterology, 120, 1666-1679.

In: CACO-2 Cells and their Uses
Editor: Megan A. Schulz, pp. 77-88

ISBN 978-61209-917-0

Chapter 5

CACO-2 CELLS AS AN EXPERIMENTAL MODEL FOR THE STUDY OF CELL-MATRIX INTERACTIONS AND WOUND HEALING IN INTESTINAL VILLUS CELLS

Amira Seltana, Nuria Basora and Jean-François Beaulieu*

CIHR Team on the Digestive Epithelium, Département d'anatomie et de biologie cellulaire, Faculté de médecine et des sciences de la santé, Université de Sherbrooke, QC, Canada, J1H 5N4

ABSTRACT

The human colon carcinoma cell line CACO-2 has been widely characterized and used to investigate intestinal cell function and barrier regulation. These cells spontaneously undertake a complete intestinal-like program of epithelial differentiation, but their cancerous nature, as well as their colonic origin, has lead to reservations as to the physiological relevance of these cells. However, numerous studies have demonstrated that CACO-2 cells share many characteristics with the differentiated villus cell population of the normal intestine. This review focuses on the use of CACO-2 cells for the study of cell-matrix interactions and wound healing processes in intestinal cells. Rapid and efficient healing of epithelial damage is critical to the functional integrity of the small intestine. Several studies have reported the ability of CACO-2 cells to synthesize and secrete extracellular components that can affect the capacity and rate of epithelial restitution. Moreover, epithelial repair is a complex process relying on the coordinated action between epithelial cells and the underlying contractile cells. We have described a fully polarized CACO-2 cell monolayer grown on top of human intestinal myofibroblasts as a co-culture system allowing the deposition of a physiologically relevant basement membrane–like structure resulting from the mutual inductive effects of each cell type on the other. Monocultures of either cell type do not generate a basement membrane, but are only capable of forming a primitive insoluble matrix. Monitoring of

*Corresponding Author: Prof. Jean-François Beaulieu, Département d'anatomie et de biologie cellulaire, Faculté de médecine et des sciences de la santé, 3001 12th Ave. N, Sherbrooke, Quebec, Canada J1H 5N4, Phone: (819) 564-5269, e-mail: Jean-Francois. Beaulieu@ USherbrooke.ca

wound healing showed drastic morphological changes in epithelial polarized columnar cells at the edge of the wound and repair was significantly more efficient in CACO-2/myofibroblast co-cultures. Taken together, these results indicate that this co-culture system is a valid tool to investigate the mechanisms underlying human intestinal epithelial restitution as well as the key players involved in their regulation.

Keywords: CACO-2, basement membrane, epithelial restitution, myofibroblasts, co-culture, wound healing

INTRODUCTION

The epithelium of the small intestine is a dynamic and highly regulated system. Cell renewal, migration and expression of digestive/absorptive abilities along the crypt-villus axis are orchestrated by a series of regulatory mechanisms that are beginning to be better understood [1-3]. Although the combination of individual intestinal normal cell models such as the human intestinal epithelial crypt (HIEC) [4] cells and primary cultures of differentiated epithelial (PCDE) [5] cells allows some recapitulation of the crypt-villus axis *in vitro* [6, 7], it is noteworthy that until very recently [8], normal human intestinal epithelial cell models able to proliferate under culture conditions and to undertake normal human epithelial differentiation have not been available.

Select transformed or cancerous cell lines have allowed insight into particular aspects of intestinal cell function and regulation [9]. However, the human colon carcinoma cell line CACO-2 is the only cell model known to be able to spontaneously express key characteristics associated with cells of the normal human intestinal epithelium, particularly in terms of the expression of polarization and differentiation markers [10]. Indeed, CACO-2 cells have been used widely and advantageously for the study of various aspects of human intestinal cell functions, namely brush border enzyme biosynthesis and processing, intracellular transport and regulation [11-16]. Many of these studies were performed on the stable clone CACO-2/15 derived from the original CACO-2 cells [13]. In our laboratory, we have been using CACO-2/15 cells at fixed passage number (between passages 53 to 85) from frozen stocks in order to minimise variability over time [14].

The unique ability of CACO-2 cells to undergo proliferation arrest accompanied by a gradual and complete enterocyte-like differentiation program at post-confluence has been well documented over the last three decades. In fact, evidence that fully differentiated CACO-2 cells compare well with villus cells of the intact small intestine has been amply provided in terms of biochemical [13-15] and ultrastructural features [14], cell-adhesion molecules [17, 18], signalization [19-21], lipid metabolism [22, 23], oxidative stress [24, 25] as well as their proteomic and genetic profiles [26, 27].

CACO-2 CELLS AND CELL-MATRIX INTERACTIONS

Because of its well-organized functional compartments containing the stem, proliferative and differentiating cells in the crypts and fully functional cells on the villi, the human intestinal epithelium has been used for the characterization of cell-matrix interactions in relation to cell state [28-30]. Indeed, differential expression of both basement membrane (BM) macromolecules such as type IV collagens, laminins, fibronectin, tenascin and SPARC [31-38] and receptors of the integrin family such as integrins α2β1, α3β1, α5β1, α7β1, α8β1, α9β1 and α6β4 [28, 39-44] has been observed.

The use of CACO-2/15 cells has led to significant progress in our understanding of the role of integrin-matrix interactions on the regulation of intestinal cell functions. Integrin α7β1, a specific laminin receptor, is restricted to the differentiating and newly differentiated cell populations *in situ* and expression has been shown to closely parallel the acquisition of differentiation characteristics in CACO-2/15 cells [43]. Integrin α7β1 was first identified as a laminin-111 receptor, and was later shown to display an affinity for laminins -211/-221 and -511 [45]. We have also shown that members of the laminin family can modulate CACO-2/15 cell differentiation-related gene expression. For example, 4C7-purified laminin, now named laminin-511, plays a key role in the regulation of functional differentiation in the enterocytic-like CACO-2/15 cell line [31]. In contrast, integrin α9β1, which is confined to the crypt compartment of the human intestine, has been found to be inversely regulated in relation to the differentiation state of CACO-2 cells. Treatment with epidermal growth factor, known to maintain CACO-2 cells in a proliferative state, results in higher levels of α9 expression, implying a relationship of α9β1 with intestinal cell proliferation [41, 46]. Additional studies using the CACO-2/15 cell model allowed us to identify the existence of various forms of the α6β4 integrin, which are functionally distinct and differentially expressed in relation to cell state under both normal and pathological conditions [42]. These findings showed that the α6Aβ4ctd- integrin form (which appears to be pro-proliferative but not functional for adhesion on purified laminin-332, one of the preferred ligands for α6β4) is expressed in normal proliferative intestinal cells while the α6Bβ4ctd+ form (anti-proliferative and functional for laminin-332 adhesion) is associated with quiescent and differentiated intestinal cells [47]. In CACO-2 cells, as well as other colon cancer cell lines, the α6Aβ4ctd+ form (an abnormal form that is pro-proliferative and functional for laminin adhesion) was found to be associated with cell proliferation while differentiated CACO-2 cells express the same form found in normal villus cells [48, 49].

The use of CACO-2 cells has also contributed to our understanding of mechanistic events involved in the healing process of the intestinal epithelium. Two major modes of regulation have been reported to have a significant impact on the maintenance of intestinal epithelial homeostasis: epithelial restitution and epithelial regeneration [50-52]. The restitution process is independent of cell proliferation and is characterized by a sequential series of events in response to loss of continuity in the epithelial monolayer. Polarized columnar cells shouldering the wound dedifferentiate, extend membrane projections to migrate across the epithelial breach and once these cells have spanned the wound area, they ultimately repolarize [53, 54]. A polarized intestinal CACO-2 cell monolayer has been widely exploited in wound healing assays as an *in vitro* model that allows simple manipulation of the external environment and easy visualization. Use of these cells grown as monolayers cultivated on plastic or an artificial coating rich in extracellular matrix (ECM) has clearly shown that intestinal restitution is actively modulated to different degrees by a variety of factors. These factors include cytokines, growth factors, ECM components, cytoskeletal rearrangements and intracellular signalling events [45, 55-60]. Recently, we described an intriguing role for integrin-linked kinase, ILK, a scaffolding protein in functional focal adhesion points, in the control of intestinal homeostasis. ILK knockdown was shown to affect CACO-2/15 cell spreading, migration, and restitution ability [61, 62], phenomena related to a decrease in fibronectin deposition [61].

Until now, epithelial cell monolayers have been considered to give an accurate reflection of events involved in wound healing dynamics within the functional crypt-villus axis [54, 63-66]. However, we and others have questioned the physiological relevance of these models and the difficulty of relating results obtained *in vitro* to what occurs *in vivo* [50, 59, 67, 68]. Previous studies have shown increased migration rates when CACO-2 cells were grown on matrices composed of collagen or fibronectin [55, 57, 60]. In our experiments, the addition of either of these two components in their soluble form to the culture medium did not significantly stimulate migration of CACO-2/15 cells during restitution when grown on plastic [59]. Recent intestinal epithelial restitution studies have highlighted the limits of this approach. For example, it was shown that single cell defects sealed ten times faster in the native colon than in HT-29 cell monolayers, emphasizing the necessity for models that better recapitulate the epithelial–stromal interface [68].

CACO-2 Cells and Cell-Matrix Interactions in Co-Culture

Superficial denuding of the intestinal epithelium occurs regularly during the normal course of digestion and absorption of a meal and must be rapidly restored. In the small intestine, the process of repair is aided on a number of levels by the underlying myofibroblasts, which reside immediately beneath the epithelial BM [69, 70]. For example, myofibroblasts mediate villus contraction, which rapidly reduces the surface area requiring repair [67]. These subepithelial cells also produce a wide variety of cytokines and growth factors [70, 71], such as transforming growth factor-β and epidermal growth factor, which directly enhance epithelial cell migration and most likely contribute to the increased rates of wound healing observed *in vivo*. Intestinal epithelial cell lines such as CACO-2/15, HT29 and T84 synthesize and secrete several ECM components and deposit them in the form of an insoluble, primitive matrix [31, 72, 73]. Experiments from our group have shown that CACO-2/15 change their profile of BM molecule expression when grown on confluent myofibroblastic cells. Indeed, key components found in the intact intestinal BM including type IV collagen (α1[IV]2α2[IV]), laminin-332, fibronectin and tenascin were all detected at the base of the epithelial layer in the presence of an underlying stromal layer composed of human intestinal myofibroblasts (HIM) [32, 34, 35, 37, 59]. These proteins were not only synthesized *de novo* but were secreted and correctly deposited in the BM located at the epithelial-stromal interface. The functionality of this BM-like structure was supported, for example, by the observed redistribution of heparan sulfate proteoglycan and the β1 integrin subunit to the basal membrane [32].

Teller et al. reported the expression and deposition of various laminin α-chains by CACO-2/15 cells in co-culture with myofibroblasts [35]. Similar to the *in situ* situation where both laminin-511/521 and 332 are detected at the base of the differentiated villus cells in the normal human intestine [74, 75], expression of the laminin α5 (laminin-511/521) and laminin α3 (laminin-332) chains were consistently detected at the epithelial-stromal interface in the CACO-2/15-HIM co-cultures while the crypt form laminin α2 (laminin-211/221) [75, 76] was not detected in this villus-like model. Transmission electron microscopy carried out on the

CACO-2/15-HIM co-culture model confirmed that, as in the native intestinal epithelium, epithelial cells formed a simple polarized epithelium that displayed functional staining for cytoskeleton and cell-cell junction components and densely packed microvilli at the apical surface. These cells sat on a stratified myofibroblast layer and a BM was clearly visible as an electron-dense structure between the two cell sheets. The functional integrity of our co-culture model was supported by the specific presence of type II hemidesmosomes [59]. Staining for sucrase-isomaltase showed the typical heterogeneous apical expression pattern found in functionally differentiated epithelial cells. Similarly, the differentiated status of myofibroblasts was confirmed by positive α-SMA detection throughout the stromal layer [59]. Overall these observations confirm that CACO-2 cells can be used as the epithelial component in a co-culture model, faithfully reflecting several key characteristics of the human intestinal epithelium and can be a useful model for the study of regulatory functions in the intestine [32, 35, 59].

CACO-2 Cells in Co-Culture: A Wound Healing Model

We have successfully adapted the CACO-2/15-HIM model to study epithelial-matrix interactions during the epithelial restitution process. To mimic this phenomenon, we created an artificial micro-wound in a fully polarized CACO-2/15 cell monolayer without disturbing the underlying stromal cell support as described by Seltana et al [59]. This approach allowed the visualization of key events and characteristics of epithelial restitution. At the wound edge, epithelial cells lost their polarity, dedifferentiated and became flattened in the typical shape adopted by migratory cells, while lateral E-cadherin was preserved (Fig.1). Interestingly, the interaction alone between the two cell types grown as a co-culture system was advantageous for healing. We showed that CACO-2/15 cells resting on a myofibroblast cell network repaired defects approximately two times faster than when grown on plastic, highlighting the mutual inductive effect and the importance of dialogue between epithelial and stromal cells in an intestinal restitution context [59].

The addition of soluble type I collagen or fibronectin further accelerated wound resealing under co-culture conditions. In our co-culture model, laminin-511/521, commonly produced by both cell types, and type IV collagen and laminin-332, produced by mesencyhmal and epithelial cells, respectively, were strongly detected at the migration front in the extended cells covering the wound area. These observations allowed us to propose a model in which epithelial cells do not only use the existing BM structure but rely on the active contribution of both cell types to regenerate during the healing process.

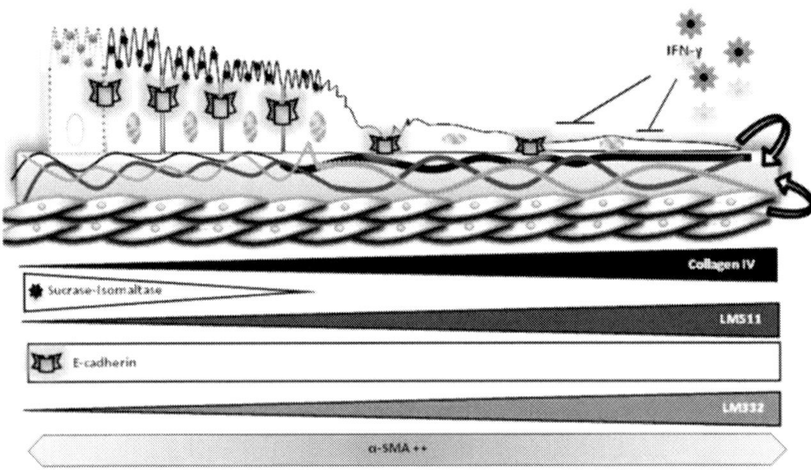

Figure 1. A model for epithelial restitution: CACO-2/15-human intestinal myofibroblast (HIM) co-cultures. Wounding induced the migration of epithelial cells on the underlying HIM monolayer, which was positive for α-SMA, a myofibroblast differentiation marker. CACO-2/15 cells underwent morphological changes and the progressive loss of a functional brush border associated with decreased expression of the differentiation marker sucrase–isomaltase. Interestingly, we identified a strong increase in expression and deposition of the basement membrane components produced by either one or both cell types. These basement membrane components were concentrated at the base of the migratory cells and likely actively contributed to epithelial restitution. Our co-culture model allowed us to distinguish between the effects of specific cytokines on cell migration in an epithelial restitution context.

The epithelial lining of the intestine forms a barrier against the external environment. Repeated damage of the intestinal surface is a crucial incident in inflammatory bowel diseases (IBD) like Crohn's disease and ulcerative colitis [77]. We have previously shown that the specific laminin α-chains, α3 and α5, are abnormally expressed and distributed along the crypt-villus axis in the small intestine of patients with Crohn's disease [74]. Moreover, Francoeur et al. subsequently showed that two proinflammatory cytokines, tumor necrosis factor-α and interferon-γ, were mainly responsible for these alterations [69, 78]. In IBD, proinflammatory cytokines have been shown to aggravate these epithelial disorders by reducing epithelial barrier function [79, 80]. A recent study has shown that single-cell restitution in HT-29 cell monolayers and native colon epithelia are differently affected by proinflammatory cytokines [68]. In our CACO-2/15-HIM co-culture model, we studied the impact of proinflammatory cytokines on the process of epithelial restitution. As reported for the native epithelium, our results clearly distinguished between an inhibitory effect of interferon-γ but not of tumor necrosis factor-α on wound repair (A. Seltana and J.F. Beaulieu, unpublished data). In contrast, studies carried out on cultured cell monolayers have shown a potentiating cell migratory effect of interferon-γ and restitution inhibition in response to tumor necrosis factor-α [68].

This further emphasizes that restitution processes in cultured cells and in native epithelia differ with respect to the underlying mechanism in response to ECM components, growth factors, and proinflammatory cytokines. Growing CACO-2/15 cells in a dynamic co-culture system that faithfully recapitulates the differentiated compartment of the small intestine allows a better understanding of intestinal regulatory functions, such as epithelial restitution.

CONCLUSIONS

Over the past decades, the CACO-2 cell line has been proven to possess the ability to stop proliferation and undertake a normal-like program of intestinal cell differentiation. This unique characteristic has allowed these cells to be used as an excellent tool to study human intestinal cell functions and the mechanisms that regulate them. However, as scientific knowledge accumulates we are increasingly aware of the limitations of single cell type models cultured on plastic to accurately reproduce the responses of the native organ and to investigate their regulation. Albeit their cancerous origin, post-confluent quiescent CACO-2/15 cells lying on top of a stromal support such as that provided by confluent human intestinal myofibroblasts appears to recapitulate to a significant extent native intestinal tissue organization. Such a co-culture system is relatively easy to reproduce and appears to be a valuable *in vitro* model for the analysis of the restitution mechanisms of the human intestinal epithelium and represents a physiologically relevant context for the characterization of the key players in this process.

ACKNOWLEDGMENTS

The authors thank Elizabeth Herring for proofreading the manuscript.

The work was supported by a grant from the Canadian Institute for Health Research MOP-97836. JFB holds the Canada Research Chair in Intestinal Physiopathology and is a member of the FRSQ-funded Centre de recherche Clinique Etienne-LeBel of the CHUS.

REFERENCES

[1] Babyatsky MW; Podolsky DK. *Growth and development of the gastrointestinal tract.* In: Yamada T, editor, editor. Textbook of gastroenterology. 3rd ed. Philadelphia: J.B Lippincott; 1999. p. 547-84.

[2] Ménard D; Beaulieu J-F; Boudreau F; Perreault N; Rivard N; Vachon PH. *Gastrointestinal tract.* In: Krieglstein KUaK, editor. Cell Siganling and Growth Factors in Development: From molecules to Organogenesis. Weinheim: Wiley-Vch:; 2006. p. 755-90.

[3] Montgomery RK; Mulberg AE; Grand RJ. *Development of the human gastrointestinal tract: twenty years of progress.* Gastroenterology. 1999, 116, 702-31.

[4] Perreault N; Beaulieu JF. *Use of the dissociating enzyme thermolysin to generate viable human normal intestinal epithelial cell cultures.* ExpCell Res. 1996, 224, 354-64.

[5] Perreault N; Beaulieu JF. *Primary cultures of fully differentiated and pure human intestinal epithelial cells.* Exp Cell Res. 1998, 245, 34-42.

[6] Pageot LP; Perreault N; Basora N; Francoeur C; Magny P; Beaulieu JF. *Human cell models to study small intestinal functions: recapitulation of the crypt-villus axis.* Microsc Res Tech. 2000, 49, 394-406.

[7] Escaffit F; Perreault N; Jean D; Francoeur C; Herring E; Rancourt C, et al. *Repressed E-cadherin expression in the lower crypt of human small intestine: a cell marker of functional relevance.* Exp Cell Res. 2005, 302, 206-20.

[8] Benoit YD; Pare F; Francoeur C; Jean D; Tremblay E; Boudreau F, et al. *Cooperation between HNF-1alpha, Cdx2, and GATA-4 in initiating an enterocytic differentiation program in a normal human intestinal epithelial progenitor cell line.* Am J Physiol Gastrointest Liver Physiol. 2010, 298, G504-17.

[9] Zweibaum A; Chantret I. *Human colon carcinoma cell lines as in vitro models for the study of intestinal cell differentiation.* In: Smith MW, Sepulveda SV, editors. Adaptation and development of gastrointestinal function. Manchester: Manchester University Press; 1989. p. 103-12.

[10] Rousset M. *The human colon carcinoma cell lines HT-29 and CACO-2: two in vitro models for the study of intestinal differentiation.* Biochimie. 1986, 68, 1035-40.

[11] Hauri HP; Sterchi EE; Bienz D; Fransen JA; Marxer A. *Expression and intracellular transport of microvillus membrane hydrolases in human intestinal epithelial cells.* J Cell Biol. 1985, 101, 838-51.

[12] Beaulieu JF; Nichols B; Quaroni A. *Posttranslational regulation of sucrase-isomaltase expression in intestinal crypt and villus cells.* J Biol Chem. 1989, 264, 20000-11.

[13] Beaulieu JF; Quaroni A. *Clonal analysis of sucrase-isomaltase expression in the human colon adenocarcinoma CACO-2 cells.* Biochem J. 1991, 280 (Pt 3), 599-608.

[14] Vachon PH; Beaulieu JF. *Transient mosaic patterns of morphological and functional differentiation in the CACO-2 cell line.* Gastroenterology. 1992, 103, 414-23.

[15] Vachon PH; Perreault N; Magny P; Beaulieu JF. *Uncoordinated, transient mosaic patterns of intestinal hydrolase expression in differentiating human enterocytes.* J Cell Physiol. 1996, 166, 198-207.

[16] Quaroni A; Paul EC; Nichols BL. *Intracellular degradation and reduced cell-surface expression of sucrase-isomaltase in heat-shocked CACO-2 cells.* Biochem J. 1993, 292 (Pt 3), 725-34.

[17] Basora N; Tetreault MP; Boucher MP; Herring E; Beaulieu JF. *Polycystin-1 is a microtubule-driven desmosome-associated component in polarized epithelial cells.* Exp Cell Res. 2010, 316, 1454-64.

[18] Escaffit F; Boudreau F; Beaulieu JF. *Differential expression of claudin-2 along the human intestine: Implication of GATA-4 in the maintenance of claudin-2 in differentiating cells.* J Cell Physiol. 2005, 203, 15-26.

[19] Dufour G; Demers MJ; Gagne D; Dydensborg AB; Teller IC; Bouchard V, et al. *Human intestinal epithelial cell survival and anoikis. Differentiation state-distinct regulation and roles of protein kinase B/Akt isoforms.* J Biol Chem. 2004, 279, 44113-22.

[20] Vachon PH; Harnois C; Grenier A; Dufour G; Bouchard V; Han J, et al. *Differentiation state-selective roles of p38 isoforms in human intestinal epithelial cell anoikis.* Gastroenterology. 2002, 123, 1980-91.

[21] Aliaga JC; Deschenes C; Beaulieu JF; Calvo EL; Rivard N. *Requirement of the MAP kinase cascade for cell cycle progression and differentiation of human intestinal cells.* Am J Physiol. 1999, 277, G631-41.

[22] Levy E; Menard D; Delvin E; Montoudis A; Beaulieu JF; Mailhot G, et al. *Localization, function and regulation of the two intestinal fatty acid-binding protein types*. Histochem Cell Biol. 2009, 132, 351-67.

[23] Sane AT; Sinnett D; Delvin E; Bendayan M; Marcil V; Menard D, et al. *Localization and role of NPC1L1 in cholesterol absorption in human intestine*. J Lipid Res. 2006, 47, 2112-20.

[24] Marcil V; Seidman E; Sinnett D; Boudreau F; Gendron FP; Beaulieu JF, et al. *Modification in oxidative stress, inflammation and lipoprotein assembly in response to hepatocyte nuclear factor 4 alpha knockdown in intestinal epithelial cells*. J Biol Chem. 2010.

[25] Taha R; Seidman E; Mailhot G; Boudreau F; Gendron FP; Beaulieu JF, et al. *Oxidative stress and mitochondrial functions in the intestinal CACO-2/15 cell line*. PLoS One. 2010, 5, e11817.

[26] Pshezhetsky AV; Fedjaev M; Ashmarina L; Mazur A; Budman L; Sinnett D, et al. *Subcellular proteomics of cell differentiation: quantitative analysis of the plasma membrane proteome of CACO-2 cells*. Proteomics. 2007, 7, 2201-15.

[27] Tremblay E; Auclair J; Delvin E; Levy E; Menard D; Pshezhetsky AV, et al. *Gene expression profiles of normal proliferating and differentiating human intestinal epithelial cells: a comparison with the CACO-2 cell model*. J Cell Biochem. 2006, 99, 1175-86.

[28] Lussier C; Basora N; Bouatrouss Y; Beaulieu JF. *Integrins as mediators of epithelial cell-matrix interactions in the human small intestinal mucosa*. Microsc Res Tech. 2000, 51, 169-78.

[29] Teller IC; Beaulieu JF. *Interactions between laminin and epithelial cells in intestinal health and disease*. Expert Rev Mol Med. 2001, 3, 1-18.

[30] Beaulieu JF. *Extracellular matrix components and integrins in relationship to human intestinal epithelial cell differentiation*. Prog Histochem Cytochem. 1997, 31, 1-78.

[31] Vachon PH; Beaulieu JF. *Extracellular heterotrimeric laminin promotes differentiation in human enterocytes*. Am J Physiol. 1995, 268, G857-67.

[32] Vachon PH; Durand J; Beaulieu JF. *Basement membrane formation and re-distribution of the beta 1 integrins in a human intestinal co-culture system*. Anat Rec. 1993, 235, 567-76.

[33] Beaulieu JF; Vachon PH; Chartrand S. *Immunolocalization of extracellular matrix components during organogenesis in the human small intestine*. Anat Embryol (Berl). 1991, 183, 363-9.

[34] Desloges N; Simoneau A; Jutras S; Beaulieu JF. *Tenascin may not be required for intestinal villus development*. Int J Dev Biol. 1994, 38, 737-9.

[35] Teller IC; Auclair J; Herring E; Gauthier R; Menard D; Beaulieu JF. *Laminins in the developing and adult human small intestine: relation with the functional absorptive unit*. Dev Dyn. 2007, 236, 1980-90.

[36] Beaulieu JF; Vachon PH; Herring-Gillam FE; Simoneau A; Perreault N; Asselin C, et al. *Expression of the alpha-5(IV) collagen chain in the fetal human small intestine*. Gastroenterology. 1994, 107, 957-67.

[37] Simoneau A; Herring-Gillam FE; Vachon PH; Perreault N; Basora N; Bouatrouss Y, et al. *Identification, distribution, and tissular origin of the alpha5(IV) and alpha6(IV) collagen chains in the developing human intestine*. Dev Dyn. 1998, 212, 437-47.

[38] Lussier C; Sodek J; Beaulieu JF. *Expression of SPARC/osteonectin/BM4O in the human gut: predominance in the stroma of the remodeling distal intestine.* J Cell Biochem. 2001, 81, 463-76.

[39] Perreault N; Vachon PH; Beaulieu JF. *Appearance and distribution of laminin A chain isoforms and integrin alpha 2, alpha 3, alpha 6, beta 1, and beta 4 subunits in the developing human small intestinal mucosa.* Anat Rec. 1995, 242, 242-50.

[40] Beaulieu JF. *Differential expression of the VLA family of integrins along the crypt-villus axis in the human small intestine.* J Cell Sci. 1992, 102 (Pt 3), 427-36.

[41] Basora N; Desloges N; Chang Q; Bouatrouss Y; Gosselin J; Poisson J, et al. *Expression of the alpha9beta1 integrin in human colonic epithelial cells: resurgence of the fetal phenotype in a subset of colon cancers and adenocarcinoma cell lines.* Int J Cancer. 1998, 75, 738-43.

[42] Basora N; Herring-Gillam FE; Boudreau F; Perreault N; Pageot LP; Simoneau M, et al. *Expression of functionally distinct variants of the beta(4)A integrin subunit in relation to the differentiation state in human intestinal cells.* J Biol Chem. 1999, 274, 29819-25.

[43] Basora N; Vachon PH; Herring-Gillam FE; Perreault N; Beaulieu JF. *Relation between integrin alpha7Bbeta1 expression in human intestinal cells and enterocytic differentiation.* Gastroenterology. 1997, 113, 1510-21.

[44] Benoit YD; Lussier C; Ducharme PA; Sivret S; Schnapp LM; Basora N, et al. *Integrin alpha8beta1 regulates adhesion, migration and proliferation of human intestinal crypt cells via a predominant RhoA/ROCK-dependent mechanism.* Biol Cell. 2009, 101, 695-708.

[45] Nishiuchi R; Takagi J; Hayashi M; Ido H; Yagi Y; Sanzen N, et al. *Ligand-binding specificities of laminin-binding integrins: a comprehensive survey of laminin-integrin interactions using recombinant alpha3beta1, alpha6beta1, alpha7beta1 and alpha6beta4 integrins.* Matrix Biol. 2006, 25, 189-97.

[46] Desloges N; Basora N; Perreault N; Bouatrouss Y; Sheppard D; Beaulieu JF. *Regulated expression of the integrin alpha9beta1 in the epithelium of the developing human gut and in intestinal cell lines: relation with cell proliferation.* J Cell Biochem. 1998, 71, 536-45.

[47] Dydensborg AB; Teller IC; Basora N; Groulx JF; Auclair J; Francoeur C, et al. *Differential expression of the integrins alpha6Abeta4 and alpha6Bbeta4 along the crypt-villus axis in the human small intestine.* Histochem Cell Biol. 2009, 131, 531-6.

[48] Dydensborg AB; Teller IC; Groulx JF; Basora N; Pare F; Herring E, et al. *Integrin alpha6Bbeta4 inhibits colon cancer cell proliferation and c-Myc activity.* BMC Cancer. 2009, 9, 223.

[49] Beaulieu JF. *Integrin alpha6 beta4 in colorectal cancer.* World J Gastrointest Pathophysiol. 2010, 1, 3-11.

[50] Blumberg RS; Li L; Nusrat A; Parkos CA; Rubin DC; Carrington JL. *Recent insights into the integration of the intestinal epithelium within the mucosal environment in health and disease.* Mucosal Immunol. 2008, 1, 330-4.

[51] Laukoetter MG; Nava P; Nusrat A. *Role of the intestinal barrier in inflammatory bowel disease.* World J Gastroenterol. 2008, 14, 401-7.

[52] Thompson JS; Saxena SK; Sharp JG. *Regulation of intestinal regeneration: new insights.* Microsc Res Tech. 2000, 51, 129-37.

[53] Lotz MM; Rabinovitz I; Mercurio AM. *Intestinal restitution: progression of actin cytoskeleton rearrangements and integrin function in a model of epithelial wound healing.* Am J Pathol. 2000, 156, 985-96.

[54] Nusrat A; Delp C; Madara JL. *Intestinal epithelial restitution. Characterization of a cell culture model and mapping of cytoskeletal elements in migrating cells.* J Clin Invest. 1992, 89, 1501-11.

[55] Basson MD. *In vitro evidence for matrix regulation of intestinal epithelial biology during mucosal healing.* Life Sci. 2001, 69, 3005-18.

[56] Basson MD; Modlin IM; Flynn SD; Jena BP; Madri JA. *Independent modulation of enterocyte migration and proliferation by growth factors, matrix proteins, and pharmacologic agents in an in vitro model of mucosal healing.* Surgery. 1992, 112, 299-307; discussion -8.

[57] Basson MD; Modlin IM; Madri JA. *Human enterocyte (CACO-2) migration is modulated in vitro by extracellular matrix composition and epidermal growth factor.* J Clin Invest. 1992, 90, 15-23.

[58] Russo JM; Florian P; Shen L; Graham WV; Tretiakova MS; Gitter AH, et al. *Distinct temporal-spatial roles for rho kinase and myosin light chain kinase in epithelial purse-string wound closure.* Gastroenterology. 2005, 128, 987-1001.

[59] Seltana A; Basora N; Beaulieu JF. *Intestinal epithelial wound healing assay in an epithelial-mesenchymal co-culture system.* Wound Repair Regen. 2010, 18, 114-22.

[60] Zhang J; Owen CR; Sanders MA; Turner JR; Basson MD. *The motogenic effects of cyclic mechanical strain on intestinal epithelial monolayer wound closure are matrix dependent.* Gastroenterology. 2006, 131, 1179-89.

[61] Gagne D; Groulx JF; Benoit YD; Basora N; Herring E; Vachon PH, et al. *Integrin-linked kinase regulates migration and proliferation of human intestinal cells under a fibronectin-dependent mechanism.* J Cell Physiol. 2010, 222, 387-400.

[62] Yuan L; Sanders MA; Basson MD. *ILK mediates the effects of strain on intestinal epithelial wound closure.* Am J Physiol Cell Physiol. 2010, (in press).

[63] Miller TA; Smith GS; Banan A; Kokoska ER. *Cytoskeleton as a target for injury in damaged intestinal epithelium.* Microsc Res Tech. 2000, 51, 149-55.

[64] Yu CF; Basson MD. *Matrix-specific FAK and MAPK reorganization during CACO-2 cell motility.* Microsc Res Tech. 2000, 51, 191-203.

[65] Zhang J; Li W; Sanders MA; Sumpio BE; Panja A; Basson MD. *Regulation of the intestinal epithelial response to cyclic strain by extracellular matrix proteins.* Faseb J. 2003, 17, 926-8.

[66] Zhang J; Li W; Sumpio BE; Basson MD. *Fibronectin blocks p38 and jnk activation by cyclic strain in CACO-2 cells.* Biochem Biophys Res Commun. 2003, 306, 746-9.

[67] Blikslager AT; Moeser AJ; Gookin JL; Jones SL; Odle J. *Restoration of barrier function in injured intestinal mucosa.* Physiol Rev. 2007, 87, 545-64.

[68] Gunzel D; Florian P; Richter JF; Troeger H; Schulzke JD; Fromm M, et al. *Restitution of single-cell defects in the mouse colon epithelium differs from that of cultured cells.* Am J Physiol Regul Integr Comp Physiol. 2006, 290, R1496-507.

[69] Francoeur C; Bouatrouss Y; Seltana A; Pinchuk IV; Vachon PH; Powell DW, et al. *Degeneration of the Pericryptal Myofibroblast Sheath by Proinflammatory Cytokines in Inflammatory Bowel Diseases.* Gastroenterology. 2008.

[70] Powell DW; Mifflin RC; Valentich JD; Crowe SE; Saada JI; West AB. *Myofibroblasts. II. Intestinal subepithelial myofibroblasts*. Am J Physiol. 1999, 277, C183-201.

[71] Powell DW; Mifflin RC; Valentich JD; Crowe SE; Saada JI; West AB. *Myofibroblasts. I. Paracrine cells important in health and disease.* Am J Physiol. 1999, 277, C1-9.

[72] Lotz MM; Nusrat A; Madara JL; Ezzell R; Wewer UM; Mercurio AM. *Intestinal epithelial restitution. Involvement of specific laminin isoforms and integrin laminin receptors in wound closure of a transformed model epithelium*. Am J Pathol. 1997, 150, 747-60.

[73] Vachon PH; Simoneau A; Herring-Gillam FE; Beaulieu JF. *Cellular fibronectin expression is down-regulated at the mRNA level in differentiating human intestinal epithelial cells.* Exp Cell Res. 1995, 216, 30-4.

[74] Bouatrouss Y; Herring-Gillam FE; Gosselin J; Poisson J; Beaulieu JF. *Altered expression of laminins in Crohn's disease small intestinal mucosa*. Am J Pathol. 2000, 156, 45-50.

[75] Beaulieu JF; Vachon PH. *Reciprocal expression of laminin A-chain isoforms along the crypt-villus axis in the human small intestine.* Gastroenterology. 1994, 106, 829-39.

[76] Simon-Assmann P; Duclos B; Orian-Rousseau V; Arnold C; Mathelin C; Engvall E, et al. *Differential expression of laminin isoforms and alpha 6-beta 4 integrin subunits in the developing human and mouse intestine.* Dev Dyn. 1994, 201, 71-85.

[77] Sturm A; Dignass AU. *Epithelial restitution and wound healing in inflammatory bowel disease*. World J Gastroenterol. 2008, 14, 348-53.

[78] Francoeur C; Escaffit F; Vachon PH; Beaulieu JF. *Proinflammatory cytokines TNF-alpha and IFN-gamma alter laminin expression under an apoptosis-independent mechanism in human intestinal epithelial cells.* Am J Physiol Gastrointest Liver Physiol. 2004, 287, G592-8.

[79] Andoh A; Bamba S; Brittan M; Fujiyama Y; Wright NA. *Role of intestinal subepithelial myofibroblasts in inflammation and regenerative response in the gut.* Pharmacol Ther. 2007, 114, 94-106.

[80] Capaldo CT; Nusrat A. *Cytokine regulation of tight junctions.* Biochim Biophys Acta. 2008.

In: CACO-2 Cells and their Uses
Editor: Megan A. Schulz, pp. 89-99

ISBN 978-61209-917-0

Chapter 6

EFFECT OF POLYUNSATURATED FATTY ACIDS AND BILE SALTS ON BUTYRATE UPTAKE BY INTESTINAL EPITHELIAL CELLS

Pedro Gonçalves, João R. Araújo and Fátima Martel*
Department of Biochemistry (U38-FCT), Faculty of Medicine,
University of Porto, Porto, Portugal

ABSTRACT

Our aim was to investigate the effect of *n*-3 PUFAs (docosahexaenoic acid (DHA) and eicosapentaenoic acid (EPA)), *n*-6 PUFAs (linoleic acid (LA), γ-linolenic acid (LNA) and arachidonic acid (AA)), conjugated linoleic acid (CLA) and the bile salts deoxycholic acid (DCA) and taurodeoxycholic acid (TDCA) upon uptake of ^{14}C-butyrate (^{14}C-BT) by intestinal epithelial cells.

CACO-2 cells were exposed to LA, LNA, AA, DHA, EPA, CLA and DCA (100 μM) for 1 to 7 days. The only significant effects were slight (7-18%) increases in ^{14}C-BT uptake found with EPA, CLA, AA and DCA. IEC-6 cells were exposed to LA, LNA, AA, DHA, EPA, CLA, DCA and TCDA (100 μM) for 2 days. LNA, EPA, CLA and DHA caused small (9-16%) decreases in ^{14}C-BT uptake. For CLA, this effect was associated with a decrease in cellular viability.

A discrete interference of *n*-3 and *n*-6 PUFAs, CLA and unconjugated bile acids with the apical uptake of BT was observed.

Keywords: bile salts; butyrate uptake; CACO-2 cells; IEC-6 cells; polyunsaturated fatty acids.

INTRODUCTION

Colorectal cancer (CRC) is one of the most common solid tumours worldwide, being the second leading cause of cancer death among men and women combined in the United States,

*Corresponding author: F. Martel, Department of Biochemistry, Faculty of Medicine of Porto, University of Porto, 4200-319 Porto, Portugal, Tel.: 351 22 5513624, Fax: 351 22 5513624, email: fmartel@med.up.pt

Australia and Europe, and a major problem in many other countries. The causes of CRC are multifactorial, but an association between reduced risk of CRC and a diet high in fruit, vegetables and fiber has been well established in epidemiological studies (Martínez *et al.* 2008).

Butyrate (BT), a short-chain fatty acid produced by anaerobic bacterial fermentation of dietary fibre within the human colon, plays a key role in colonic epithelium homeostasis. One of the proposed beneficial effects of BT on human colonic health is the prevention/inhibition of colon carcinogenesis (Hamer *et al.* 2008). Many epidemiological studies show an inverse relationship between dietary fibre intake and the incidence of CRC (Martínez *et al.* 2008), and exposure of many colon tumour cell lines to BT leads to anticarcinogenic effects by induction of cell differentiation and apoptosis and by inhibition of proliferation (Heerdt *et al.*1994; Hague *et al.* 1995). At the molecular level, the anticarcinogenic effect of BT is considered to depend on regulation of gene expression, which is mainly attributed to its inhibition of histone deacetylase, resulting in hyperacetylation of histones and enhancement of the accessibility of transcription factors to nucleosomal DNA (Heerdt *et al.*1994; Hague *et al.* 1995).

BT is transported into colonic epithelial cells by two specific carrier-mediated transport systems, the monocarboxylate transporter 1 (MCT1) and the sodium monocarboxylate transporter 1 (SMCT1) (Gupta *et al.* 2006; Morris and Felmlee 2008). Interestingly enough, both MCT1 and SMCT1 were recently proposed to function as tumour suppressors (Cuff *et al.* 2005; Gupta *et al.* 2006). So, interference with either MCT1- and SMCT1-mediated transport of BT into colonocytes will obviously have a major impact in the context of colon carcinogenesis. In this context, MCT1-mediated transport at the intestinal level is known to be up-regulated by its substrate, BT (Cuff *et al.* 2002), enhanced by leptin (Buyse *et al.* 2002), phorbol 12-myristate 13-acetate (Alrefai *et al.* 2004), caffeine and acetylsalicylic acid (Gonçalves *et al.* 2009), and inhibited by enteropathogenic *E. coli* (Borthakur *et al.* 2006), interferon-γ, tumour necrosis factor-α (Thibault *et al.* 2007), theophylline, tetrahydro cannabinol, MDMA (ecstasy), acetaldehyde and indomethacin (Gonçalves *et al.* 2009), and by some polyphenols (Konishi *et al.* 2003; Vaidyanathan and Walle 2003; Wang and Morris 2007; Shim *et al.* 2007).

n-3 (n-3) and n-6 (n-6) polyunsaturated fatty acids (PUFAs) are essential fatty acids necessary for human health. The most widely studied effects of PUFAs are those related to eicosanoid biosynthesis and function. In general terms, eicosanoids derived from n-6 PUFAs have pro-inflammatory whereas those derived from n-3 precursors have anti-inflammatory effects. Likewise, eicosanoids derived from these two series have opposing effects in cancer cell growth, invasion, and angiogenesis (Berquin et al. 2008; Chapkin et al. 2008). More specifically, in relation to CRC, and from a dietary perspective, a growing body of experimental, epidemiological and clinical evidence supports the contention that fish oil and its bioactive food components (n−3 PUFAs (eg. eicosapentaenoic acid (EPA) and docosahexaenoic acid (DHA)) are important in suppressing CRC (although several clinical trials and epidemiological studies are inconsistent), and that the opposite effect is observed with n-6 PUFAs (eg., linoleic acid (LA) and arachidonic acid (AA)), which are found in meat and vegetable oils (Berquin et al. 2008; Chapkin et al. 2008).

Conjugated linoleic acid (CLA) is a collective term for isomers of LA that have conjugated double bonds. CLA has been shown to have a number of health promoting actions, including an *in vitro* and *in vivo* anticarcinogenic effect. Among the several mechanisms proposed to be involved in its anticarcinogenic effect are alteration of lipid peroxidation, tissue fatty acid composition, eicosanoid metabolism and gene expression (Kelley *et al.* 2007).

On the other hand, secondary (unconjugated) bile acids, including deoxycholic acid (DCA), but not primary bile acids, have been shown to be tumour promoters, both *in vitro* and in animal studies (Reddy *et al.* 1977; Mahmoud *et al.* 1999). Moreover, raised levels of secondary bile acids have been reported in patients with adenomatous polyps and colon cancer (Reddy and Wynder 1977; Imray *et al.* 1992). Their mechanism of action is still poorly understood, although PKC, MAP kinases and NFκB have been identified as molecular targets (Huang *et al.* 1992; Pongracz *et al.* 1995; Payne *et al.* 2007).

Interestingly enough, the effects of BT and either PUFA (Hofmanová et al. 2005, 2009), CLA (Nichenametla et al. 2004), or unconjugated bile acids (McMillan et al. 2000; Rosignoli et al. 2008) upon colonic epithelial cell population may be mutually influenced. So, the aim of this study was to investigate the hypothesis that interference with the apical uptake of BT may be one of the mechanisms contributing to the effect of n-3 PUFA, n-6 PUFA, CLA and unconjugated bile acids upon CRC carcinogenesis. For this, we investigated the effect of n-3 PUFAs (DHA and EPA), n-6 PUFAs (LA, γ-linolenic acid (LNA) and AA), CLA, the unconjugated bile acid DCA and the conjugated bile acid taurochenodeoxycholic acid (TCDA) upon uptake of 14C-BT by a human colonic adenocarcinoma cell line (Caco-cells) and by a rat small intestinal epithelial cell line (IEC-6 cells). CACO-2 cells derive from a human colon adenocarcinoma, but when fully differentiated they form a polarized monolayer of cells expressing a combination of colonocyte and enterocyte phenotypes (Sambuy et al. 2005). On the other hand, IEC-6 cells were established from crypt of rat intestinal cells, but when grown in post-confluent culture, they develop structural changes and differentiation from a crypt cell-like to an enterocyte-like phenotype (Wood et al. 2003). Comparison between the effect of these compounds upon BT uptake in carcinogenic and noncarcinogenic cell lines seemed interesting in the context of the distinct effect of BT in these cells (Hamer et al. 2008).

Materials and Methods

CACO-2 Cell Culture

The CACO-2 cell line was obtained from the Deutsche Sammlung von Mikroorganismen and Zellkulturen (Braunschweig, Germany) and was used between passage number 51-64. The cells were maintained in a humidified atmosphere of 5% CO_2-95% air and were grown in Minimum Essential Medium (Sigma, St. Louis, MO, USA) containing 5.55 mM glucose and supplemented with 15% fetal calf serum, 25 mM HEPES, 100 units/ml penicillin, 100 μg/ml streptomycin and 0.25 μg/ml amphotericin B (all from Sigma). Culture medium was changed every 2- 3 days and the culture was split every 7 days. For sub-culturing, the cells were removed enzymatically (0.25% trypsin-EDTA, 5 min, 37°C), split 1:3, and sub-cultured in plastic culture dishes (21-cm^2; ∅ 60 mm; Corning Costar, Corning, NY, USA). For uptake studies, CACO-2 cells were seeded on 24-well plastic cell culture clusters (2-cm^2; ∅ 16 mm;

Corning Costar), and ^{14}C-BT uptake experiments were performed 7 days after the initial seeding. For 24 h before the experiments, the cell medium was made free of fetal calf serum.

IEC-6 Cell Culture

The IEC-6 cell line was obtained from Deutsche Sammlung von Mikroorganismen und Zellkulturen (ACC-111, Braunschweig, Germany) and was used between passage number 35-37. The cells were maintained in a humidified atmosphere of 5% CO_2-95% air and were cultured in Dulbecco's Modified Eagle's Medium:RPMI 1640 medium (1:1), supplemented with 10% fetal bovine serum, 0.1 U/ml insulin, 5.96 g HEPES, 2.2 g $NaHCO_3$, 100 units/ml penicillin, 100 μg/ml streptomycin and 0.25 μg/ml amphotericin B (all from Sigma). Culture medium was changed every 2 to 3 days and the culture was split every 7 days. For sub-culturing, the cells were removed enzymatically (0.05% trypsin-EDTA, 5 min, 37°C), split 1:3, and sub-cultured in plastic culture dishes (21 cm^2; Ø 60 mm; Corning Costar, Corning, NY, USA). For uptake studies, IEC-6 cells were seeded on 24-well plastic cell culture clusters (2 cm^2 Ø 16 mm; Corning Costar), and ^{14}C-BT uptake experiments were performed 9 days after the initial seeding (90-100% confluence). For 24 h before the experiments, the cell medium was made free of fetal calf serum and insulin.

Determination of ^{14}C-BT uptake by CACO-2 and IEC-6 Cells

The uptake experiments were performed with CACO-2 and IEC-6 cells incubated in glucose-free Krebs buffer (containing, in mM: 125 NaCl, 4.8 KCl, 1.2 $MgSO_4$, 1.2 $CaCl_2$, 25 NaHCO3, 1.6 KH_2PO_4, 0.4 K_2HPO_4, and 20 MES (pH 6.5)). Initially, the culture medium was aspirated and the cells were washed twice with 0.5 ml buffer at 37°C. Then, uptake was initiated by the addition of 0.3 ml medium at 37°C containing ^{14}C-BT 10 μM. Incubation was stopped after 3 min by removing the incubation medium, placing the cells on ice and rinsing the cells with 0.5 ml ice-cold buffer. The cells were then solubilised with 0.3 ml 0.1% (v/v) Triton X-100 (in 5 mM Tris.HCl, pH 7.4), and placed at 37°C overnight. Radioactivity in the cells was measured by liquid scintillation counting.

Treatment of the Cells with Compounds

The effect of compounds on ^{14}C-BT uptake was tested by cultivating cells in culture medium containing these compounds (or the respective solvent) for 1, 2, 3 or 7 days. The medium was renewed daily, and the end of the treatment period was day 7 and 9 of CACO-2 and IEC-6 cell culture, respectively.

Determination of Cell Viability

To test whether the compounds affected cellular viability, the MTT (3-(4,5-dimethyl-2-thiazolyl)-2,5-diphenyl-2H-tetrazolium bromide) assay was performed. In brief, cells were treated with the compounds as described above, and at the end of that period, the MTT assay was done as described elsewhere (Gonçalves *et al.* 2009).

Protein Determination

The protein content of cell monolayers was determined as described elsewhere (Bradford 1976), using human serum albumin as standard.

Calculation and Statistics

Arithmetic means are given with SEM, and geometric means are given with 95% confidence limits. Statistical significance of the difference between two groups was evaluated by the Student's t-test. Differences were considered to be significant when $P<0.05$.

Materials

^{14}C-BT (*n*-butyric acid, sodium salt, [1-^{14}C]; specific activity 30-60 mCi/mmol (Biotrend Chemikalien GmbH, Köln, Germany); MES (2-[N-morpholino]ethanesulfonic acid) hydrate), arachidonic acid, docosahexaenoic acid, linoleic acid, γ-linolenic acid, eicosapentahenoic acid, conjugated linoleic acid (mixture of cis- and trans-9,11- and -10,12-octadecadienoic acids), deoxycholic acid, MTT (3-(4,5-dimethyl-2-thiazolyl)-2,5-diphenyl-2H-tetrazolium bromide), taurodeoxycholic acid, trypsin-EDTA solution (Sigma, St. Louis, MO, USA); DMSO (dimethylsulfoxide), triton X-100 (Merck, Darmstadt, Germany); fetal calf serum (Invitrogen Corporation, Carlsbad, CA, USA).

Drugs to be tested were dissolved in ethanol or DMSO, the final concentration of the solvents being 1% in the culture media. Controls were run in the presence of the solvent.

Results

Previously, our group have shown that ^{14}C-BT apical uptake in CACO-2 and IEC-6 cells was found to be linear with time for up to 3 min of incubation (Gonçalves et al. 2009, 2010). So, in the present work, cells were incubated with ^{14}C-BT for 3 min, in order to determine initial rates of uptake.

Effect of Pufas and Bile Acids on ^{14}C-BT Apical uptake by CACO-2 Cells

The effect of several distinct PUFAs on ^{14}C-BT apical uptake by CACO-2 cells was investigated in the first series of experiments. For this, 100 μM of each fatty acid was tested over different periods (1, 2, 3 and 7 days). Of the fatty acids tested, LA, LNA acid and DHA were devoid of significant effect (results not shown; *n*=6-9). However, AA, EPA and CLA caused small but significant increases in ^{14}C-BT uptake at some of the time points analysed (Fig. 1A).

In the next series of experiments, we further investigated the effect of AA and EPA, by characterizing its concentration-dependency. When tested for 2 days, and contrary to what was found with 100 µM, higher concentrations of AA and EPA decreased uptake of ^{14}C-BT (1 mM of AA or EPA reduced uptake by 20%; Fig. 1B). This decrease was associated with a decrease in cellular viability (to 65.6±1.7 and 78.6±0.8% of control, respectively; *n*=10). Interestingly enough, treatment with AA (300 µM) for 7 days, which caused no decrease in ^{14}C-BT uptake (Fig. 1B), was associated with a 24.3±6.6% increase in CACO-2 viability (*n*=10). None of these treatments (1 mM of AA or EPA for 2 days or 300 µM AA for 7 days) affected CACO-2 cell proliferation (results not shown).

The effect of the bile acid DCA (100 µM) was also investigated. Apart from a small increase found after 2 days of treatment (to 109.4±2.5% of control; *n*=9), DCA was devoid of significant effect on ^{14}C-BT uptake (results not shown).

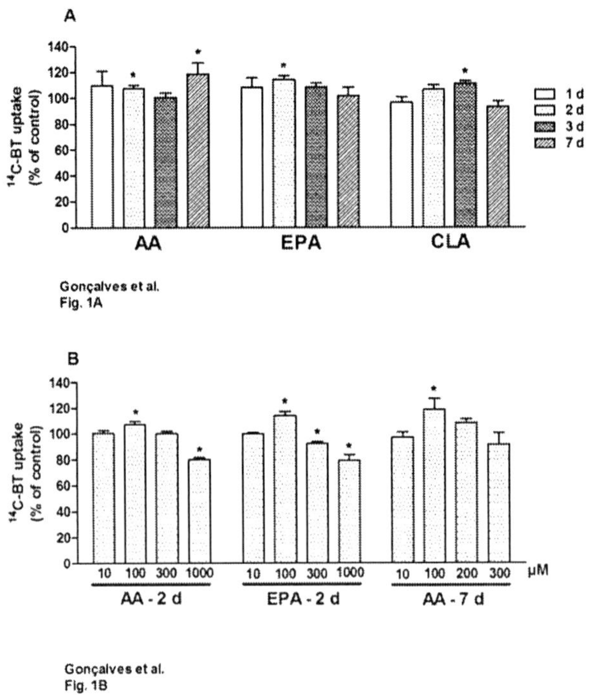

Figure 1. Effect of PUFA on 14C-BT uptake by CACO-2 cells. A. Time-dependence of the effect of PUFAs on the apical uptake of 14C-BT. Initial rates of uptake were determined in CACO-2 cells incubated for 3 min with 10 µM 14C-BT after treatment for 1, 2, 3 or 7 days with arachidonic acid 100 µM (AA; n=9), eicosapentaenoic acid 100 µM (EPA; n=9-15) or conjugated linoleic acid 100 µM (CLA; n=9-13); B. Concentration-dependency of the effect of PUFAs on the apical uptake of 14C-BT. Initial rates of uptake were determined in CACO-2 cells incubated for 3 min with 10 µM 14C-BT after treatment for 2 days with increasing concentrations of arachidonic acid or eicosapentaenoic acid (AA - 2 days and EPA - 2 days, respectively; n=9) or after treatment for 7 days with increasing concentrations of arachidonic acid (AA; n=7-9). Shown are arithmetic means±SEM. * significantly different from control.

Effect of Pufas and Bile Acids on ^{14}C-BT Apical uptake by IEC-6 Cells

We then investigated the effect of distinct PUFAs on ^{14}C-BT uptake by IEC-6 cells. For this, IEC-6 cells were treated with 100 μM of each fatty acid for 2 days. Of the compounds tested, LA and AA were devoid of significant effect (results not shown; *n*=9). On the other hand, LNA, EPA, CLA and DHA caused small (9-16%) but significant decreases in ^{14}C-BT uptake (Fig. 2A). The effect of LNA and EPA upon ^{14}C-BT uptake was found to be concentration-dependent, but this was not observed with DHA (Fig. 2B).

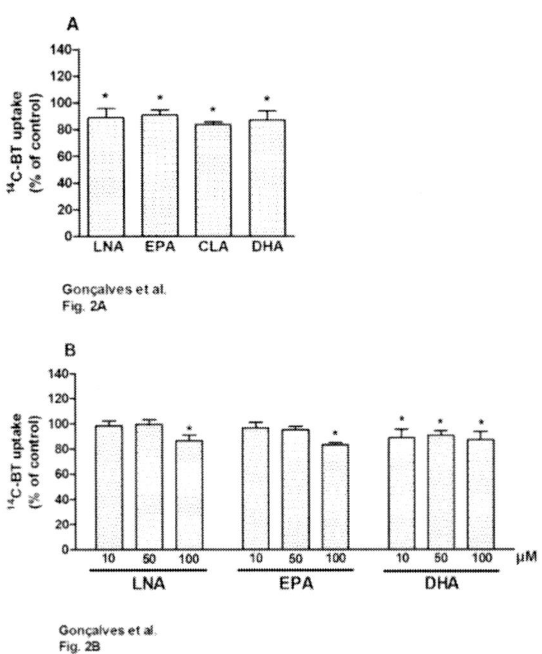

Figure 2. Effect of PUFAs on the apical uptake of 14C-BT by IEC-6 cells. A. Initial rates of uptake were determined in IEC-6 cells incubated for 3 min with 10 μM 14C-BT after treatment for 2 days with γ-linolenic acid 100 μM (LNA; n=6), eicosapentaenoic acid 100 μM (EPA; n=9), conjugated linoleic acid 100 μM (CLA; n=9), or docosaenoic acid (DHA; n=9); B. Concentration-dependency of the effect of PUFAs on the apical uptake of 14C-BT. Initial rates of uptake were determined in IEC-6 cells incubated for 3 min with 10 μM 14C-BT after treatment for 2 days with increasing concentrations of γ-linolenic acid (LNA; n=6-9), eicosapentaenoic acid (EPA; n=9) or docosahexaenoic acid (DHA; n=3-9). Shown are arithmetic means±SEM. * significantly different from control.

Interestingly enough, the inhibitory effect of 100 μM LNA, EPA and DHA was not associated with a decrease in cellular viability, as these compounds increased cellular viability (by 14.8±1.1, 21.8±1.1 and 21.9±2.4%, respectively; *n*=8-9). For CLA, however, the decrease in ^{14}C-BT uptake was associated with a similar decrease in cell viability (to 89.7±0.4% of control; *n*=8).

Finally, the effect of a 2-day exposure to the bile acids DCA and TCDA (100 μM) was also investigated. None of the compounds affected ^{14}C-BT uptake by IEC-6 cells (results not shown).

CONCLUSION

The majority of studies in the literature concerning the effects of BT, PUFAs, CLA and bile acids in colon carcinogenesis have considered these dietary factors individually. *In vivo*, however, they are commonly present together in the colon and so they may influence each other's actions directly or indirectly. For this reason, we decided to test the effect of *n*-3 PUFAs (DHA and EPA), *n*-6 PUFAs (LA, LNA and AA), CLA and the bile acids DCA and TDCA upon uptake of ^{14}C-BT by a human colonic adenocarcinoma (CACO-2) cell line and by a rat small intestinal epithelial (IEC-6) cell line. BT plays an essential role in the maintenance of colonic tissue homeostasis, and one of its proposed beneficial effects is the prevention/inhibition of colon carcinogenesis (Heerdt *et al.* 1994; Hague *et al.* 1995). So, understanding of the regulation of its absorption by the colon mucosa appears to be particularly important in the context of carcinogenesis.

Exposure of CACO-2 cells (from 1 to 7 days) to LA, LNA and DHA (100 μM) did not affect ^{14}C-BT uptake. In contrast, exposure to 100 μM EPA and DCA (for 2 days), CLA (for 3 days) or AA (for 2 and 7 days) caused small (7-18%) but significant increases in ^{14}C-BT uptake. Treatment of the cells for 2 days with higher concentrations of either AA or EPA (1000 μM) caused a 20% decrease in ^{14}C-BT uptake, which was however associated with a parallel decrease in cellular viability. These latter results suggest that the reduction of ^{14}C-BT uptake found with the higher concentrations of either AA or EPA is related to cell death. Further support for this conclusion was the observation that treatment with a high concentration of AA (300 μM) for 7 days did not affect ^{14}C-BT uptake, and was associated with a 25% increase in CACO-2 viability.

Exposure of IEC-6 cells for 2 days to LA, AA, DCA and TDCA (100 μM) did not affect ^{14}C-BT uptake. In contrast, exposure to LNA, EPA, CLA or DHA (100 μM) caused small (9-16%) but significant decreases in ^{14}C-BT uptake. The decrease in ^{14}C-BT uptake obtained with CLA (100 μM) was associated with a parallel decrease in cellular viability. However, LNA, EPA and DHA decreased ^{14}C-BT uptake while increasing cellular viability.

In summary, ^{14}C-BT uptake is modestly modulated by some PUFA and bile salts, independently of an effect on cell viability. In CACO-2 cells, EPA, AA, CLA and DCA (100 μM) increased uptake, and in IEC-6 cells LNA, EPA and DHA (100 μM) decreased it.

In CACO-2 cells, ^{14}C-BT uptake was slightly increased by compounds possessing either anticarcinogenic (EPA, CLA) or carcinogenic (AA, DCA) activity against CRC. The same was observed in IEC-6 cells, in which ^{14}C-BT uptake was slightly decreased by compounds possessing either anticarcinogenic (EPA, DHA) or carcinogenic (LNA) activity against CRC. This strongly suggests that modulation of ^{14}C-BT uptake is not involved in the effect of these compounds on CRC. In other words, given the anticarcinogenic effect of BT in CRC cells, the increase in ^{14}C-BT uptake in CACO-2 cells caused by EPA and CLA can be one of the mechanisms contributing to their anticarcinogenic effect against CRC, but cannot contribute to the carcinogenic effect of AA and DCA. On the other hand, in IEC-6 cells, the decrease in ^{14}C-BT uptake observed with LNA can contribute to the carcinogenic effect of this PUFA in relation to CRC, but cannot contribute to the anticarcinogenic effect of EPA and DHA.

We conclude that a discrete Interference of *n*-3 and *n*-6 PUFAs, CLA and unconjugated bile acids with the apical uptake of BT was observed. So, this does not seem to be one of the main mechanisms involved in the recognized role that these compounds have in colon carcinogenesis.

ACKNOWLEDGMENTS

This work was supported by FCT and Programa Ciência, Tecnologia e Inovação do Quadro Comunitário de Apoio (PTDC/SAU-FCF/67805/2006).

REFERENCES

Alrefai, W.A., Tyagi, S., Gill, R.K., Saksena S, Hadjiagapiou, C, Mansour, F., Ramaswamy, K. and Dudeja, P.K. 2004. *Regulation of butyrate uptake in CACO-2 cells by phorbol 12-myristate 13-acetate*. Am. J. Physiol. Gastrointest. Liver Physiol. *286*, 197-203.

Berquin, I.M., Edwards, I.J. and Chen, Y.Q. 2008. *Multi-targeted therapy of cancer by omega-3 fatty acids*. Cancer Lett. *269*, 363–377.

Borthakur, A., Gill, R.K., Hodges, K., Ramaswamy, K., Hecht, G. and Dudeja, P.K. 2006. *Enteropathogenic Escherichia coli inhibits butyrate uptake in CACO-2 cells by altering the apical membrane MCT1 level*. Am. J. Physiol. Gastrointest. Liver Physiol. *290*, G30-G35.

Bradford, M.M. 1976. *A rapid method for the quantitation of microgram quantities of protein utilizing the principle of protein-dye binding*. Anal. Biochem. *72*, 248-54.

Buyse, M., Sitaraman, S.V., Liu, X., Bado, A. and Merlin, D. 2002. *Luminal leptin enhances CD147/MCT-1-mediated uptake of butyrate in the human intestinal cell line Caco2-BBE*. J. Biol. Chem. *277*, 28182-28190.

Chapkin, R.S., Seo, J., McMurray, D.N. and Lupton, J.R. 2008. *Mechanisms by which docosahexaenoic acid and related fatty acids reduce colon cancer risk and inflammatory disorders of the intestine*. Chem. Phys. Lipids. *153*, 14–23.

Cuff, M., Dyer, J., Jones, M. and Shirazi-Beechey, S. 2005. *The human colonic monocarboxylate transporter isoform 1: its potential importance to colonic tissue homeostasis*. Gastroenterology *128*, 676-686.

Cuff, M.A., Lambert, D.W. and Shirazi-Beechey, S.P. 2002. *Substrate-induced regulation of the human colonic monocarboxylate transporter, MCT1*. J. Physiol. *539*, 361-371.

Gonçalves, P., Araújo, J.R. and Martel, F. 2010. *Comparison of butyrate uptake by two non-transformed intestinal epithelial cell lines*. J. Physiol. Biochem. (in press)

Gonçalves, P., Araújo, J.R., Pinho, M.J. and Martel, F. 2009. *Modulation of butyrate transport in CACO-2 cells. Naunyn-Schmiedeberg's Arch.* Pharmacol. *379*, 325-336.

Gupta, N., Martin, P.M., Prasad, P.D. and Ganapathy, V. 2006. *SLC5A8 (SMCT1)-mediated transport of butyrate forms the basis for the tumor suppressive function of the transporter*. Life Sci. *78*, 2419-2425.

Hague, A., Elder, D.J., Hicks, D.J. and Paraskeva, C. 1995. *Apoptosis in colorectal tumour cells: induction by the short chain fatty acids butyrate, propionate and acetate and by the bile salt deoxycholate*. Int. J. Cancer. *60*, 400-406.

Hamer, H.M., Jonkers, D., Venema, K., Vanhoutvin, S., Troost, F.J. and Brummer, R.J. 2008. *Review article: the role of butyrate on colonic function*. Aliment. Pharmacol. Ther. *27*, 104-119.

Heerdt, B.G., Houston, M.A. and Augenlicht, L.H. 1994. *Potentiation by specific short-chain fatty acids of differentiation and apoptosis in human colonic carcinoma cell lines*. Cancer Res. *54*, 3288-3293.

Hofmanová, J., Vaculová, A., Koubková, Z., Hyzd'alová, M., Kozubík, A. 2009. *Human fetal colon cells and colon cancer cells respond differently to butyrate ans PUFAs*. Mol. Nutr. Food Res. *53*, S102-S113.

Hofmanová, J., Vaculová, A., Lojek, A. and Kozubík, A. 2005. *Interaction of polyunsaturated fatty acids and sodium butyrate during apoptosis in HT-29 human colon adenocarcinoma cells*. Eur. J. Nutr. *44*, 40–51.

Huang, X.P., Fan, X.T., Desieux, J.F., and Castagna, M. 1992. *Bile acids, non-phorbol ester type tumor promoters stimulate the phosphorylation of protein kinase C substrates in human platelets and colon cell line HT29*. Int. J. Cancer. *52*, 444–450.

Imray, C.H.E., Radley, S., Davis, A., Barker, G., Hendrickse, C.W., Donovan, I.A., Lawson, A.M., Baker, P.R. and Neoptolemos, J.P. 1992. *Fecal unconjugated bile acids in patients with colorectal cancer or polyps*. Gut *33*, 1239–1245.

Kelley, N.S., Hubbard, N.E. and Erickson, K.L. 2007. *Conjugated linoleic acid isomers and cancer*. J. Nutr. *137*, 2599–2607.

Konishi, Y., Kobayashi, S. and Shimizu, M. 2003. *Tea polyphenols inhibit the transport of dietary phenolic acids mediated by the monocarboxylic acid transporter (MCT) in intestinal CACO-2 cell monolayers*. J. Agric. Food Chem. *51*, 7296-302.

Mahmoud, N.N., Dannenberg, A.J., Bilinski, R.T., Mestre, J.R., Chadburn, A., Churchill, M., Martucci, C and, Bertagnolli, M.M. 1999. *Administration of an unconjugated bile acid increases duodenal tumors in a murine model of familial adenomatous polyposis*. Carcinogenesis *20*, 299–303.

Martínez, M.E., Marshall, J.R. and Giovannucci, E. 2008. *Diet and cancer prevention: the roles of observation and experimentation*. Nat. Rev. Cancer *8*, 694-703.

McMillan, L., Butcher, S., Wallis, Y., Neoptolemos, J.P. and Lord, J.M. 2000. *Bile acids reduce the apoptosis-inducing effects of sodium butyrate on human colon adenoma (AA/C1) cells: Implications for colon carcinogenesis*. Biochem. Biophys. Res. Commun. *273*, 45–49.

Morris, M.E. and Felmlee, M.A. 2008. *Overview of the proton-coupled MCT (SLC16A) family of transporters: characterization, function and role in the transport of the drug of abuse gamma-hydroxybutyric acid*. AAPS J. *10*, 311-321.

Nichenametla, S.N., South, E.H. and Exon, J.H. 2004. *Interaction of conjugated linoleic acid, sphingomyelin, and butyrate on formation of colonic aberrant crypt foci and immune fuctions in rats*. J. Toxicol. Environm. Health (Part A) *67*, 469–481.

Payne, C.M., Weber, C., Crowley-Skillicorn, C., Dvorak, K., Bernstein, H., Bernstein, C., Holubec, H., Dvorakova, B. and Garewal, H. 2007. *Deoxycholate induces mitochondrial oxidative stress and activates NF-kappaB through multiple mechanisms in HCT-116 colon epithelial cells*. Carcinogenesis *28*, 215-22.

Pongracz, J., Clark, P., Neoptolemos, J.P. and Lord, J.M. 1995. *Expression of protein kinase C isoenzymes in colorectal cancer tissue and their differential activation by different bile acids*. Int. J. Cancer *61*, 35–39.

Reddy, B.S. and Wynder, E.L. 1977. *Faecal bile acids and neutral sterols in colon cancer patients and patients with adenomatous polyps.* Cancer *39*, 2533–2539.

Reddy, B.S., Watanabe, K., Weissburger, J.H. and Wynder, E.L. 1977. *Promoting effect of bile acids in colon carcinogenesis in germ-free and conventional F344 rats.* Cancer Res. *37*, 3238–3242.

Rosignoli, P., Fabiani, R., De Bartolomeo, A., Fuccelli, R., Pelli, M.A. and Morozzi, G. 2008. *Genotoxic effect of bile acids on human normal and tumour colon cells and protection by dietary antioxidants and butyrate.* Eur. J. Nutr. *47*, 301-309.

Sambuy, Y., De Angelis, I., Ranaldi, G., Scarino, M.L., Stammati, A. and Zucco, F. 2005. *The CACO-2 cell line as a model of the intestinal barrier: influence of cell and culture-related factors on CACO-2 cell functional characteristics.* Cell Biol. Toxicol. *21*, 1-26.

Shim, C.K., Cheon, E.P., Kang, K.W., Seo, K.S. and Han, H.K. 2007. *Inhibition effect of flavonoids on monocarboxylate transporter 1 (MCT1) in CACO-2 cells.* J. Pharm. Pharmacol. *59*, 1515-1519.

Thibault, R., De Coppet, P., Daly, K., Bourreille, A., Cuff, M., Bonnet, C., Mosnier, J.F., Galmiche, J.P., Shirazi-Beechey, S. and Segain, J.P. 2007. *Down-regulation of the monocarboxylate transporter 1 is involved in butyrate deficiency during intestinal inflammation.* Gastroenterology *133*, 1916-1927.

Vaidyanathan, J.B. and Walle, T. 2003. *Cellular uptake and efflux of the tea flavonoid (-)epicatechin-3-gallate in the human intestinal cell line CACO-2.* J. Pharmacol. Exp. Ther. *307*, 745-752.

Wang, Q. and Morris, M.E. 2007. *Flavonoids modulate monocarboxylate transporter-1-mediated transport of gamma-hydroxybutyrate in vitro and in vivo.* Drug Metab. Dispos. *35*, 201-208.

Wood, S.R., Zhao, Q., Smith, L.H. and Daniels, C.K. 2003. *Altered morphology in cultured rat intestinal epithelial IEC-6 cells is associated with alkaline phosphatase expression.* Tissue & Cell. *35*, 47–58.

In: CACO-2 Cells and their Uses
Editor: Megan A. Schulz, pp. 101-111
ISBN 978-61209-917-0

Chapter 7

ANALYSIS OF BOTULINUM TOXIN: INTESTINAL EPITHELIAL BARRIER INTERACTION USING EPITHELIAL MONOLAYER CULTURE SYSTEMS

Yukako Fujinaga*[*], *Takuhiro Matsumura and Yo Sugawara
Laboratory for Infection Cell Biology, International Research Center for Infectious Diseases, Research Institute for Microbial Diseases, Osaka University, 3-1 Yamada-oka, Suita, Osaka, 565-0871, Japan

ABSTRACT

Food-borne and intestinal (infant and adult) botulism are the most common forms of human botulism, and both result from the passage of botulinum neurotoxin (BoNT) from the digestive tract into the circulation. Interestingly, BoNT retains its toxicity after passage through the digestive tract despite being a large protein toxin (approximately 150 kDa). BoNTs are produced as large protein complexes (BoNT complex) associated with non-toxic components, including non-toxic non-hemagglutinin (NTNH) and hemagglutinin (HA). These proteins protect BoNT from the acidic and proteolytic environment of the digestive tract, and, moreover, we recently showed that HA disrupts the intercellular epithelial barrier by directly binding to E-cadherin. As an *in vitro* human intestinal epithelial barrier model, the CACO-2 monolayer system is among best experimental systems to advance these studies. In this chapter, we will provide an overview of the recently elucidated molecular interactions between the BoNT complex and the intestinal epithelial barrier.

INTRODUCTION

Botulinum neurotoxin (BoNT, serotypes A–G, protein toxin of MW ~150 kDa) is a bacterial toxin produced by *Clostridium botulinum* and related species. BoNT causes

[*]Corresponding author: Fujinaga, Y. (yukafuji@biken.osaka-u.ac.jp)

botulism, a severe neurological disease characterized by flaccid paralysis, and it exhibits metalloprotease activity. BoNT specifically binds neuronal cells, enters the cytoplasm, and cleaves the core proteins involved in vesicular fusion thereby blocking neurotransmitter release (reviewed in (1)). Among the seven BoNT serotypes, types A, B, E and F cause botulism in both humans and animals, but types C and D primarily cause disease in domestic animals. They are rarely associated with clinically significant disease in humans. Type G producing organisms have been isolated from only soil (reviewed in (2)).

Bacterially produced BoNT is non-covalently associated with nontoxic proteins (non-toxic neurotoxin-associated proteins, NAP). (reviewed in (2-6), Fig. 1). There are three major forms of BoNT complex isolated from bacterial cultures: 12S toxin (M toxin), 16S toxin (L toxin), and 19S toxin (LL toxin). The 12S toxin is composed of BoNT and a non-toxic non-HA component (NTNH, MW ~130 kDa). The 16S toxin is composed of a BoNT, NTNH and hemagglutinin (HA) proteins. The 19S toxin has the same components as 16S toxin, and it is thought to be a dimer of two 16S toxins linked by one of the HA proteins, HA1 (7). There are three HA proteins: HA1 (MW ~33 kDa), HA2 (MW ~17 kDa), and HA3 (MW ~70 kDa). In some bacterial strains, HA3 is cleaved into HA3a (MW ~20 kDa) and 3b (MW ~50 kDa) (reviewed in (5)). The type A strain of *C. botulinm* produces 12S, 16S and 19S toxins while the type B, C, and D strains produce 12S and 16S toxins. The type E and F strains produce only 12S toxin, and type G strain produces 16S toxin. BoNT dissociates from the nontoxic components under slightly alkaline conditions (higher than pH 7 ~ 8), but the BoNT complex remains intact in intestinal fluid despite a pH of ~ 7 (reviewed in (3)). In seminal studies in the 1970s, Sakaguchi's group found that the oral toxicity of BoNT increases with the incremental association of nontoxic components with BoNT, and they proposed that NTNH and HA confer resistance to proteolysis within the gastrointestinal tract (reviewed in (3)).

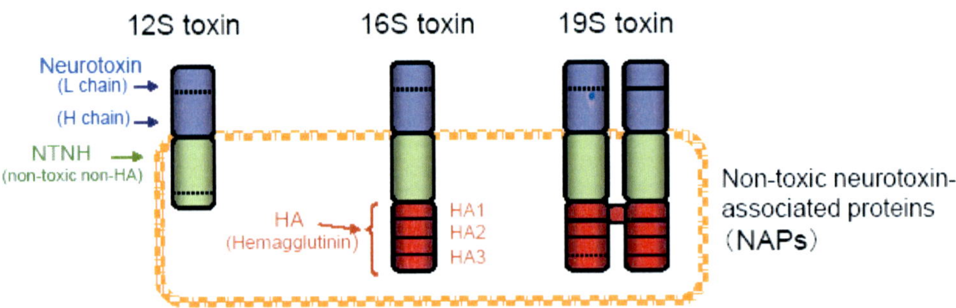

Figure 1. Schematic structure of botulinum neurotoxin (BoNT) complexes. Three major forms of BoNT complexes, 12S toxin (M toxin), 16S toxin (L toxin), and 19S toxin (LL toxin), are presented (reviewed in (3-6)). 12S toxin is composed of a BoNT and a non-toxic non-HA (NTNH). 16S toxin is composed of a BoNT, an NTNH and several hemagglutinin (HA) proteins. 19S toxin has the same components as 16S toxin, and is presumed to be a dimer of two 16S toxins linked by one of the HA proteins, HA1 (7). The HA component has three different proteins; HA1, HA2, and HA3. In some strains, HA3 is cleaved into HA3a and 3b (reviewed in (5)). Dashed line denotes the proteolytic cleavage site.

Food-borne and intestinal botulism are the primary manifestations of botulism in humans. In the former, BoNT contaminated in food causes the disease, and in the latter, BoNT-producing clostridia colonize the intestinal lumen of affected infants and adults, leading to paralysis. In these cases, intoxication develops only if BoNT is absorbed from the digestive tract into the systemic circulation and reaches the peripheral nerves. The epithelial barrier of the digestive tract is a major obstacle preventing the absorption and dissemination of BoNT. Although the molecular mechanisms by which this large protein toxin crosses the epithelial barrier are not completely defined, recent studies performed by us and others have increased the understanding of the factors involved in the interaction of BoNT and BoNT complexes with the epithelial barrier. In these studies, *in vitro* model systems simulating the human intestinal epithelial barrier are essential, and one such system is the CACO-2 monolayer system (Table 1). Here we provide an overview of the current state of knowledge of the pathway by which BoNT and BoNT complex are transported across the intestinal epithelial barrier.

Table 1. Papers describing the interaction of botulinum toxin with the intestinal epithelial barrier using cultured epithelial monolayer systems

Cell line	Serotype of botulinum toxin	Main subject or main finding	Assessment of integrity of monolayer	Papers
T-84, CACO-2, MDCK	A, B, C	Transcytosis of BoNT alone	Flux of [^{3}H] inulin Leakage of medium	J. Biol. Chem 273: 21950-21957 (1998)
T-84, MDCK	A, B	C terminal domain of H chain of BoNT mediates transcytosis.	Flux of [^{3}H] inulin	J. Phermacol. Exp. Ther. 310: 633-641 (2004)
T-84	A	Visualization of binding and transcytosis of BoNT	Measurement of TER	J. Phermacol. Exp. Ther. 315: 1028-1035 (2005)
CACO-2	D	HA1 in the BoNT complex enhances the binding and passage of BoNT through CACO-2 monolayer.	No description	FEMS Immunol. Med. Microbiol. 49 346-532 (2007)
CACO-2, MDCK	B	The HA proteins of BoNT complex disrupt intestinal epithelial intercellular junctions to increase toxin absorption.	Measurement of TER Flux of FITC-dextran Staining of junctional proteins including ZO-1	Cell. Microbiol. 10:355-364 (2008)

Table 1. (Continued).

CACO-2, T84, m-IC$_{cl2}$, MDCK, Vero, Cos-7, HeLa, PC12, NG108-15	A	Interaction of BoNT H chain with gangliosides and SV-2 (or a related protein) mediates transcytosis.	Measurement of TER Flux of FITC-dextran ZO-1 staining	Cell. Microbiol. 10, 375-387 (2008)
CACO-2, m-IC$_{cl2}$, NG108-15	A	BoNT H chain is endocytosed via a Cdc42-dependent and clathrin-independent pathway in intestinal cells.	No description	Cell. Microbiol. 11, 289-308 (2009)
CACO-2, T84, MDCK, ACL-15, RCN-9	A, B, C	Analysis of the mechanism of the barrier disrupting activity of different types of HAs in the BoNT complex.	Measurement of TER Flux of FITC-dextran	Microbiology 155: 35-45 (2009)
IEC-6	D	Interaction of NAPs of the 16S toxin with cell surface sialic acid leads to 16S passage though the IEC-16 monolayer.	No description	Vet. Microbiol. 141: 312-320 (2010)
CACO-2, MDCK, L cell	A, B, C	Type B HA directly binds epithelial cadherin (E-cadherin) and disrupts E-cadherin–mediated cell to cell adhesion and, in turn, the epithelial barrier.	Measurement of TER Staining of junctional proteins including ZO-1	J. Cell Biol. 189: 691-700 (2010)

Throughout this chapter, when available data do not differentiate between the effects of BoNT or BoNT complexes (i.e., BoNT plus NAP), we will generally refer to the toxin as 'botulinum toxin'.

1. Transcytosis of BoNT across Cultured Intestinal Epithelial Monolayers

Beginning in the late 1990s, experiments using human intestinal epithelial cell monolayers (CACO-2 and T84) showed that BoNT/A and BoNT/B bind to these cells and undergo transcytosis from the apical to the basolateral side (8-12). The heavy chain (Hc) of BoNT was identified as the domain mediating the BoNT binding and transcytosis (9, 11, 13), and intestinal cell expressed gangliosides (GD1b and GT1b) and SV2 (or SV2-related protein) were found to be involved in these processes (11). Consistent with this, a mouse intestinal crypt-like cell line (m-ICc12), which expresses higher levels of SV2C (or SV2C related protein) than CACO-2, exhibits greater binding and transcytosis of BoNT/A (11). Although the BoNT transcytosis pathway has not yet been clearly defined, it was recently reported that the C-terminus of the Hc of BoNT/A is taken up via a Cdc42-dependent and clathrin-independent pathway in CACO-2 and m-ICc12 cells, and it reaches an early endosomal compartment (13).

2. Adhesion of HA to Intestinal Epithelial Cells

The HA component demonstrates carbohydrate-binding activity that mediates the agglutination of red blood cells (reviewed in (14)). Among the three subcomponents of HA, HA1 and HA3 have carbohydrate-binding activity (reviewed in (14)). In the late 1990s, Fujinaga *et al.* reported that HA of type C 16S toxin binds to carbohydrates expressed by guinea pig intestinal epithelium, suggesting that HA facilitates the intestinal absorption of BoNT and may contribute to the high oral toxicity of the 16S toxin (15). Subsequently, HAs from other serotypes of the botulinum toxin complex were found to bind to intestinal epithelial cells via cell surface carbohydrates (16-20). Furthermore, it was reported that the type C 16S toxin, but not the 12S toxin or BoNT alone, induces toxin complex internalization and transport of BoNT to the Golgi in HT-29 cells, a human colon carcinoma cell line (21, 22). In addition, it was reported that HA1 from the type D BoNT complex enhances the cell binding and passage of BoNT across intestinal epithelial monolayers, CACO-2 and IEC-6 (a rat intestinal epithelial cell line) (18, 20). Taken together, these data suggest that HA binding to cell-expressed carbohydrates may induce the transport of BoNT across the intestinal epithelial monolayer via endocytosis and vesicular trafficking, a process referred to as transcytosis. In contrast, NAP does not facilitate the transcytosis of BoNT in T84, CACO-2, and Calu-3 cells (a human pulmonary cell line) (type A (9, 11, 23), type B (9)). Thus, the role of the carbohydrate-binding capacity of HA in mediating BoNT uptake and transcytosis remains controversial and may differ among HA serotypes.

3. Disruption of the Intercellular Epithelial Barrier by HA

Recently, we found that type B HA disrupts the intercellular epithelial barrier (CACO-2 and MDCK) (12) (Fig. 2). HA disrupts the localization of tight and adherens junction proteins, e.g. occludin, ZO-1, β-catenin and E-cadherin, at cell-to-cell boundaries without affecting the viability of cultured epithelial cells. Furthermore, we showed that HA facilitates the paracellular transport of 12S toxins in a mouse *in situ* loop assay. Both type A and type B HA proteins exhibit similar barrier-disrupting activity on epithelial monolayers derived from several species including humans, but type C HA lacks this activity for human cell lines (CACO-2 and T84) (24). However, type C HA disrupts the epithelial barrier in cultured cells derived from canines (MDCK), but this disruption was due to toxin-mediated cytotoxicity and fundamentally differed from the epithelial disruption caused by HA from type A and B toxins (24). Type C HA also shows cytotoxicity to rat-derived intestinal epithelial cell lines (ACL-15 and RCN-9) (24). At this time, the effects of type C HA-induced epithelial barrier disruption have only been examined using cell lines *in vitro*, and *in vivo* studies using susceptible animals are needed to validate its role in the pathophysiology of botulism.

Having established a role for HA in BoNT toxicity, we wished to identify the cellular factors responsible for HA mediated intercellular barrier disruption, and we determined that epithelial cadherin (E-cadherin) is a target of type B HA (Fig. 3) (25, 26). E-cadherin is a canonical member of the cadherin superfamily of adhesion molecules and mediates calcium-dependent cell-to-cell adhesion (27). In epithelial cells, E-cadherin is present in adherens junctions and is essential for the function of tight junctions (28). HA directly binds to E-cadherin and disrupts E-cadherin–mediated cell-to-cell adhesion. All three HA proteins were required for full binding to E-cadherin, although a complex consisting of only HA2 and HA3 exhibited weak binding activity. The carbohydrate-binding activity of HA is not required for this interaction. HA binds to human, bovine, and mouse E-cadherin, but not the rat or chicken homologs. Additionally, HA does not interact with other members of the classic cadherin family, such as N- and VE-cadherin. We also showed that residue 20 of E-cadherin is critically involved in its interaction with HA. This residue is located in the distal extracellular cadherin domain (EC1) near the cadherin trans-dimer interface (29).

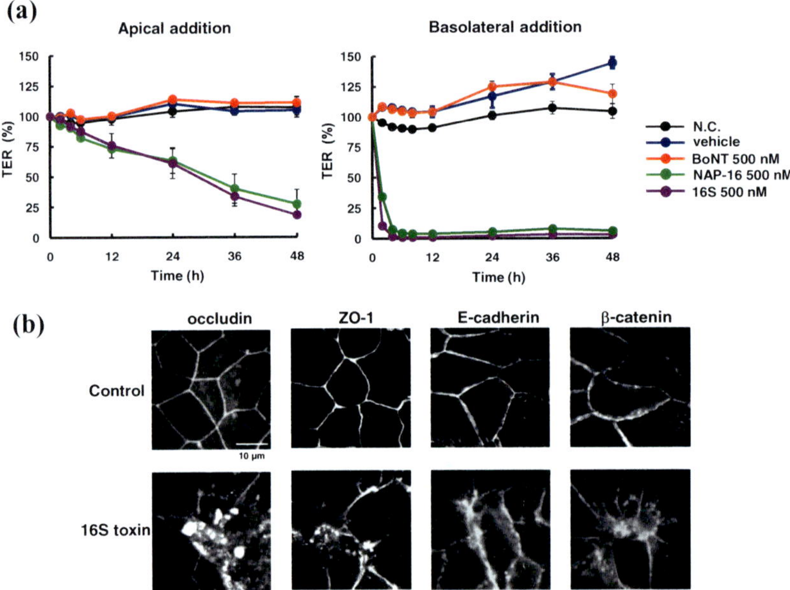

Figure 2. HA of type B BoNT complex disrupts the intercellular barrier of CACO-2 monolayer. (a) Left panel: Reduction in TER following apical (left panel) or basolateral addition (right panel) of the B 16S toxin and NAP (a complex of NTNH + HA). The indicated surface of CACO-2 monolayers were incubated without (black •) or with vehicle alone (10 mM phosphate buffer, pH 6.0) (blue •), 500 nM BoNT (red •), 500 nM 16S toxin (purple •), or 500 nM NAP-16 (green •) before measurement of TER. Values are means ± SE (n=3). (b) Altered distribution of junctional proteins following toxin treatment. CACO-2 monolayers were treated apically with vehicle (10 mM phosphate buffer, pH 6.0) (none) or 500 nM type B16S toxin for 24 h. Cells were fixed and stained for junctional proteins. Scale bar, 10 μm.

Therefore, we hypothesize that HA binding to E-cadherin sterically hinders E-cadherin trans-dimer formation. Anatomically, this presents a conundrum because E-cadherin is exclusively found on the basolateral surface of epithelial cells, and orally ingested HA in the BoNT complex is initially localized at the apical surface. Consistent with this, we showed in an *in vitro* epithelial model that the HA target site is on the basolateral surface, and HA must be transported from the apical to the basolateral surface to co-localize with E-cadherin. Finally, our data indicate that transcytosis mediates the transport of HA, at least *in vitro* (Fig. 3, (12, 25)).

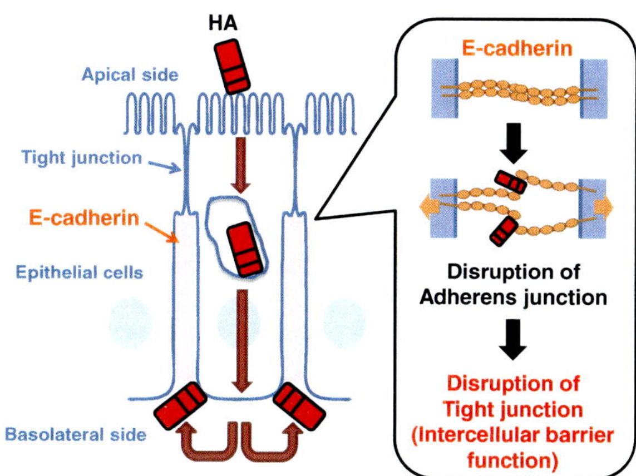

Figure 3. Schematic representation of the mechanism of action of type B HA (25). HA (HA1/2/3 complex) bound to the apical surface of epithelial cells gains access to the basolateral compartment likely via transcytosis. Translocated HA interacts with the extracellular domain of E-cadherin, leading to the disruption of adherens junctions and tight junctions, which play a vital role in the intercellular barrier function.

E-cadherin binding by HA was species-specific, and the observed species specificity correlated with the epidemiology of botulism (25). For example, type B HA did not interact with chicken E-cadherin, which is consistent with the rare reports of avian botulism caused by BoNT/B. Consistent with this, birds were experimentally resistant to the BoNT/B complex, especially when administered orally (30, 31). Additionally, BoNT/A and B complexes interact with human E-cadherin, while BoNT/C complex does not, and this correlates with the frequency of botulism caused by these types in humans (i.e., types A and B are much more common than type C) (25). Taken together, these findings suggest that the HA–E-cadherin interaction is an important factor in the pathogenesis of food-borne botulism. Additional studies using genetically modified animals expressing HA-resistant isoforms of E-cadherin are required to validate these hypotheses.

4. Possible Mechanisms of BoNT Complex Transport across the Intestinal Epithelial Monolayer

Our studies using cultured intestinal epithelial cell monolayers demonstrate that the type B HA proteins present in the 16S toxin act on E-cadherin on the basolateral surface and disrupt the intercellular epithelial barrier. Furthermore, *in situ* experiments indicate that this disruption facilitates the efficient transepithelial absorption of luminal macromolecules including 12S toxin (12). Considering these findings and those of other research groups, we propose that the botulinum toxin (serotype A and B) traverses the intestinal epithelial barrier in three steps (Figure 4). In Step 1, a small amount of luminal BoNT complexes (Route i) (12) or BoNT alone (Route ii) (8-13) undergo transcytosis across the epithelium without disrupting the epithelial intercellular barrier. Once on the basolateral surface, the HA moiety of the 16S toxin mediates disruption of the intercellular barrier [Step 2] (12, 24, 25). Finally, in Step 3, the absence of an intact intercellular barrier allows for the passage of large amounts of BoNT complexes (and BoNT) paracellularly (Route iii) (12).

The molecular mechanisms regulating the transcytosis of HA (Route i) are largely uncharacterized, and it remains unclear whether of the entire BoNT complex (16S toxin) is transported to the basolateral compartment via this route.

Future studies should determine the anatomic and cellular location of each of these transport routes (e.g., M cell or undifferentiated crypt cells). Finally and most importantly, the relevance of these cell biological phenomena to the pathogenesis of food-borne and intestinal botulism remains unestablished. This can only be addressed by identifying the important molecular interactions between botulinum toxin (BoNT, BoNT complex, and HA) and the intestinal epithelium using appropriate *in vivo* models.

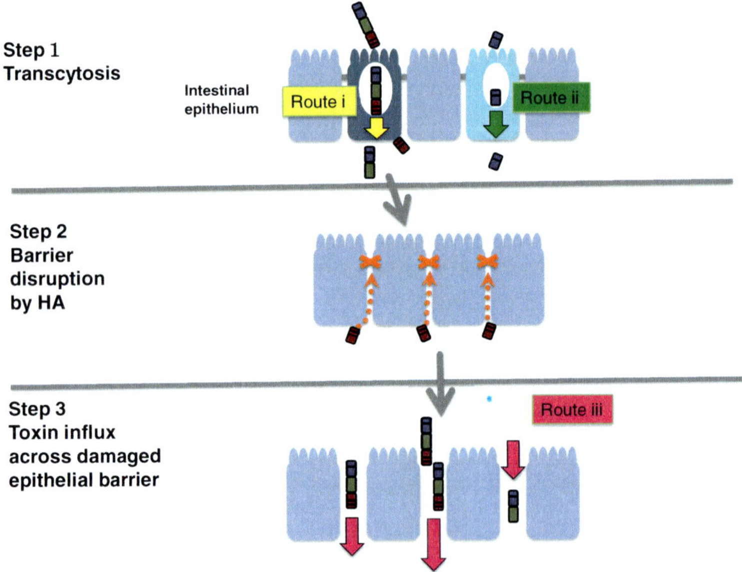

Figure 4. Proposed model for the penetration of types A and B BoNT complexes across the intestinal epithelial barrier. [Step 1] Within the BoNT complex, HA mediates the binding and transcytosis of a small amount of luminal BoNT complexes across the epithelium without disrupting the epithelial barrier (Arrows in yellow, "Route i") (12). The H chain of BoNT also mediates its binding and transcytosis (Arrows in green, "Route ii") (8-13). [Step 2] After translocation to the basolateral surface, HA disrupts the epithelial barrier (12, 24). Type A and B HA proteins bind E-cadherin and disrupt the intercellular barrier without causing cytotoxic effects in epithelial cells of susceptible hosts (25). [Step 3] A large amount of BoNT complexes and BoNT accumulate on the basolateral side by passing across the damaged epithelial intercellular barrier (Arrows in light green, "Route iii") (12).

Conclusion

In the past decade, we have made significant progress in understanding the interaction of botulinum toxin with the intestinal epithelial barrier through the use of cultured epithelial monolayer systems (Table 1). A large body of research has shown that BoNT binds to human intestinal epithelial monolayers (CACO-2 and T84) and undergoes transcytosis from the apical to basolateral side (8-13). It has also become clear that HA proteins, which have been thought to be "non-toxic", disrupt the epithelial barrier function (12, 24), and E-cadherin is the target molecule of type B HA (25). This finding indicates that HA proteins are pathogenic factors that breach host defense mechanisms by directly disrupting the epithelial barrier, promoting the intestinal absorption of BoNT during food-borne and intestinal botulism. Additional *in vitro* and *in vivo* studies using appropriate host species are needed to fully understand the mechanisms involved in BoNT absorption. These studies will provide not only insight into the molecular mechanisms regulating food-borne and intestinal botulism but also unique and powerful opportunities to understand the complicated mechanisms that maintain and regulate epithelial barriers.

Acknowledgments

Our studies described in this chapter were partly supported by grants from the Ministry of Education, Culture, Sports, Science and Technology of Japan (Y. F. and T. M.).

References

[1] Schiavo G, Matteoli M, Montecucco C. *Neurotoxins affecting neuroexocytosis.* Physiol Rev. 2000;80:717-66.
[2] Collins MD, East AK. *Phylogeny and taxonomy of the food-borne pathogen Clostridium botulinum and its neurotoxins.* J Appl Microbiol. 1998;84:5-17.
[3] Sakaguchi G. *Clostridium botulinum toxins.* Pharmacol Ther. 1982;19:165-94.
[4] Minton NP. *Molecular genetics of clostridial neurotoxins.* Curr Top Microbiol Immunol. 1995;195:161-94.
[5] Oguma K, Inoue K, Fujinaga Y, Yokota K, Watanabe T, Ohyama T, et al. *Structure and function of Clostridium botulinum progenitor toxin.* J Toxicol-Toxin Reviews. 1999;18:17-34.
[6] Poulain B, Popoff MR, Molgo J. *How do the Botulinum Neurotoxins block neurotransmitter release: from botulinum to the molecular mechanism of action.* The Botulinum J. 2008;1:14-87.
[7] Inoue K, Fujinaga Y, Watanabe T, Ohyama T, Takeshi K, Moriishi K, et al. *Molecular composition of Clostridium botulinum type A progenitor toxins.* Infect Immun. 1996;64:1589-94.
[8] Maksymowych AB, Simpson LL. *Binding and transcytosis of botulinum neurotoxin by polarized human colon carcinoma cells.* J Biol Chem. 1998;273:21950-7.
[9] Maksymowych AB, Simpson LL. *Structural features of the botulinum neurotoxin*

molecule that govern binding and transcytosis across polarized human intestinal epithelial cells. J Pharmacol Exp Ther. 2004;310:633-41.

[10] Ahsan CR, Hajnoczky G, Maksymowych AB, Simpson LL. *Visualization of binding and transcytosis of botulinum toxin by human intestinal epithelial cells.* J Pharmacol Exp Ther. 2005;315:1028-35.

[11] Couesnon A, Pereira Y, Popoff MR. *Receptor-mediated transcytosis of botulinum neurotoxin A through intestinal cell monolayers.* Cell Microbiol. 2008;10:375-87.

[12] Matsumura T, Jin Y, Kabumoto Y, Takegahara Y, Oguma K, Lencer WI, et al. *The HA proteins of botulinum toxin disrupt intestinal epithelial intercellular junctions to increase toxin absorption.* Cell Microbiol. 2008;10:355-64.

[13] Couesnon A, Shimizu T, Popoff MR. *Differential entry of botulinum neurotoxin A into neuronal and intestinal cells.* Cell Microbiol. 2009;11:289-308.

[14] Fujinaga Y. *Interaction of botulinum toxin with the epithelial barrier.* J Biomed Biotechnol. 2010;2010:974943.

[15] Fujinaga Y, Inoue K, Watanabe S, Yokota K, Hirai Y, Nagamachi E, et al. *The haemagglutinin of Clostridium botulinum type C progenitor toxin plays an essential role in binding of toxin to the epithelial cells of guinea pig small intestine, leading to the efficient absorption of the toxin.* Microbiology. 1997;143 (Pt 12):3841-7.

[16] Fujinaga Y, Inoue K, Nomura T, Sasaki J, Marvaud JC, Popoff MR, et al. *Identification and characterization of functional subunits of Clostridium botulinum type A progenitor toxin involved in binding to intestinal microvilli and erythrocytes.* FEBS Lett. 2000;467:179-83.

[17] Kojima S, Eguchi H, Ookawara T, Fujiwara N, Yasuda J, Nakagawa K, et al. *Clostridium botulinum type A progenitor toxin binds to Intestine-407 cells via N-acetyllactosamine moiety.* Biochem Biophys Res Commun. 2005;331:571-6.

[18] Niwa K, Koyama K, Inoue S, Suzuki T, Hasegawa K, Watanabe T, et al. *Role of nontoxic components of serotype D botulinum toxin complex in permeation through a CACO-2 cell monolayer, a model for intestinal epithelium.* FEMS Immunol Med Microbiol. 2007;49:346-52.

[19] Arimitsu H, Sakaguchi Y, Lee JC, Ochi S, Tsukamoto K, Yamamoto Y, et al. *Molecular properties of each subcomponent in Clostridium botulinum type B haemagglutinin complex.* Microb Pathog. 2008;45:142-9.

[20] Niwa K, Yoneyama T, Ito H, Taira M, Chikai T, Kouguchi H, et al. *Sialic acid-dependent binding and transcytosis of serotype D botulinum neurotoxin and toxin complex in rat intestinal epithelial cells.* Vet Microbiol. 2010;141:312-20.

[21] Nishikawa A, Uotsu N, Arimitsu H, Lee JC, Miura Y, Fujinaga Y, et al. *The receptor and transporter for internalization of Clostridium botulinum type C progenitor toxin into HT-29 cells.* Biochem Biophys Res Commun. 2004;319:327-33.

[22] Uotsu N, Nishikawa A, Watanabe T, Ohyama T, Tonozuka T, Sakano Y, et al. *Cell internalization and traffic pathway of Clostridium botulinum type C neurotoxin in HT-29 cells.* Biochim Biophys Acta. 2006;1763:120-8.

[23] Park JB, Simpson LL. *Inhalational poisoning by botulinum toxin and inhalation vaccination with its heavy-chain component.* Infect Immun. 2003;71:1147-54.

[24] Jin Y, Takegahara Y, Sugawara Y, Matsumura T, Fujinaga Y. *Disruption of the epithelial barrier by botulinum haemagglutinin (HA) proteins - differences in cell tropism and the mechanism of action between HA proteins of types A or B, and HA*

proteins of type C. Microbiology. 2009;155:35-45.

[25] Sugawara Y, Matsumura T, Takegahara Y, Jin Y, Tsukasaki Y, Takeichi M, et al. *Botulinum hemagglutinin disrupts the intercellular epithelial barrier by directly binding E-cadherin.* J Cell Biol. 2010;189:691-700.

[26] Sugawara Y, Fujinaga Y. *The botulinum toxin complex meets E-cadherin on the way to its destination.* Cell Adh Migr. 2010;5.

[27] Takeichi M. *The cadherins: cell-cell adhesion molecules controlling animal morphogenesis.* Development. 1988;102:639-55.

[28] Citi S. *The molecular organization of tight junctions.* J Cell Biol. 1993;121:485-9.

[29] Boggon TJ, Murray J, Chappuis-Flament S, Wong E, Gumbiner BM, Shapiro L. *C-cadherin ectodomain structure and implications for cell adhesion mechanisms.* Science. 2002;296:1308-13.

[30] Gross WB, Smith LD. *Experimental botulism in gallinaceous birds.* Avian Dis. 1971;15:716-22.

[31] Notermans S, Dufrenne J, Kozaki S. *Experimental botulism in Pekin ducks.* Avian Dis. 1980;24:658-64.

In: CACO-2 Cells and their Uses
Editor: Megan A. Schulz, pp. 113-128

ISBN 978-1-61209-917-0
2011 Nova Science Publishers, Inc.

Chapter 8

CACO-2 CELLS AND GENOTOXICITY STUDIES OF FOOD CONTAMINANTS

V. Fessard [*] ***and L. Le Hégarat***

Anses, French Agency for Food, Environmental and Occupational Health Safety,
Fougères Laboratory, Unit of Toxicology of contaminants,
La Haute Marche, Fougères cedex, FRANCE

ABSTRACT

The strategy to estimate the genotoxicity of xenobiotics includes *in vitro* assays detecting mutations or primary DNA lesions. Among the various cell lines that can be used, the intestinal ones are relevant for the study of food contaminants as the first barrier to cross to reach the systemic circulation. The human CACO-2 cells is a common model for toxicokinetics and numerous publications indicate specific activities (influx and efflux transporters, metabolic enzymes) which can play a key role in the toxicity of food contaminants. *In vitro* genotoxicity studies have been recently performed on this cell line and most of them are reported in this review. Some improvements to overcome the limits which can be encountered with the CACO-2 cell line can also be proposed depending on the aim of the study. As examples, co-cultures with other cell types as well as regulating metabolic activities can be considered.

INTRODUCTION ON *IN VITRO* GENOTOXICITY TESTING

Genotoxicity is a toxicological endpoint used to determine the potential of compounds to induce stable alterations into the cellular genome. The mutagenicity of a substance is, to some point, linked to its carcinogenicity (Brambilla et al., 2010; Sasaki et al., 2000). Conceptually, a battery of tests has been organized within a decision tree to evaluate the mutagenicity of compounds (Cimino, 2006; Parry et al, 2000). This strategy, based on both *in vitro* and *in*

[*] Anses, French Agency for Food, Environmental and Occupational Health Safety, Fougeres Laboratory, Unit of Toxicology of contaminants, La Haute Marche, BP 90203, 35 302 Fougères cedex, FRANCE, Email: valerie.fessard@anses.fr

vivo tests, enables to detect the panel of mutations reported as gene, chromosome and genome mutations. The most famous gene mutation assay is the Ames test involving the response of prokaryotes (bacteria) and their ability to reverse mutations efficiently after only few generations. The mouse lymphoma assay is also largely used to detect gene mutation in mammalian cells as well as the chromosome aberration test detecting chromosome alterations. More recently, the *in vitro* micronucleus assay on mammalian cells has been developed (Kirsch-Volders et al., 2003; Fenech, 2000). One of the advantages of this test consists in its capability for detecting both chromosome and genome mutations. Fluorescent *in situ* hybridization of centromeres can be carried out simultaneously to distinguish between each type of mutations (Kirsch-Volders et al., 1997). The *in vitro* micronucleus assay is quite simple, with easy scoring. It also gives information on the cytotoxicity of the compound as well as on its mechanism of action (Fenech, 2007). However, this assay cannot be performed on quiescent cells and is therefore restricted to cells showing some mitotic activity, eventually leading to the formation of micronuclei. After more than ten years for validation including interlaboratory testing, an OECD guideline (n°487) was finally adopted in July 2010 for the *in vitro* micronucleus assay which is supposed to replace other assays detecting chromosome mutations like the chromosome aberration test.

The strategy for detecting genotoxicity of a compound includes also *in vivo* assays. The micronucleus assay is generally performed on bone marrow or circulating erythrocytes of rodents. However, for ingested xenobiotics, the gastrointestinal tract might be one of the most relevant organs to be exposed. This is the reason why an *in vivo* micronucleus assay was developed on the colon of rodents. Compared to the bone marrow, it was shown to be more sensitive to detect food genotoxic compounds and mainly aneugenic ones (Coffing et al., 2010; Vanhauwaert et al., 2001). Therefore, the intestine is considered as a specific organ to evaluate the genotoxicity of food contaminants and the development of *in vitro* assays on intestinal models is pertinent and must be encouraged.

Some complementary assays can also be performed, both *in vitro* and *in vivo*, to understand the mechanism of action or to confirm the genotoxicity of xenobiotics. Those assays can be classified whether into direct tests detecting the formation of lesions (adduct post-labeling, comet assay,…) or indirect ones measuring the repair activity (unscheduled DNA synthesis,…).

The development of *in vitro* intestinal models to study the genotoxicity of xenobiotics is integrated in the will of reducing animal testing (restriction up to prohibition) claimed as one objective of various regulations mainly in Europe. As an example, cosmetics industries were struggled to replace animal testing for acute toxicity since 2009 including an implementation to chronic toxicity in a short timetable, for 2013 (EC 1223/2009). The REACH (Registration, Evaluation, Authorisation and Restriction of Chemical substances) regulation involves also the restriction of animal testing and, as frequently as possible, the use of alternative methods is requested. However, in most cases, development of alternative methods including *in vitro* models showing high predictivity for human risk assessment continues to be a challenge.

In Vitro Human Intestinal Cells: Pertinence for Genotoxicity Testing of Food Contaminants

In this context, one of the goals to achieve is to build up reliable *in vitro* assays, depending on the endpoint of interest and the pertinence of the model. Due to differences between non-human mammalian cells and human ones, mainly in terms of metabolism, the development of *in vitro* tests on human cell lines is of major interest. Besides its main contact with food contaminants, the intestine is also responsible for the absorption and the first metabolism of xenobiotics before they can reach the liver. Therefore, the use of human intestinal cells should be implemented for the toxicological study of ingested xenobiotics, including for genotoxicity testing.

Various cell types of human intestinal cells are available and commercialized by the American Cell Culture (ATCC). Among the most famous ones, CACO-2 and HT29 can be highlighted. Both present "organ-specific" performances, like proteolytic enzymes production, metabolic enzymes and tight junction protein expressions. Moreover, these cell lines are able to differentiate after few weeks in culture in enterocyte-like cells showing a polarized morphology (Brandon et al., 2006; Chantret et al., 1998). CACO-2 cells preferentially differentiate into enterocytes, showing a homogeneous polarized monolayer. Many papers have been published on this model including cell characterization and comparison to human enterocytes (Englund et al., 2006; Anderle et al., 2003; Engman et al., 2001).

Moreover, while most of the other cell systems commonly used for genotoxicity can be deficient in protein transporters involved in xenobiotic absorption, a large panel of influx (OATP, ...) and efflux (MDR1, MRPs) proteins are expressed in the CACO-2 cell line (Hayeshi et al., 2008; Maubon et al., 2007; Hilgenford et al., 2007; Calcagno et al., 2006; Seithel et al., 2006; Taipalensuu et al., 2001). This can be of concern for compounds requiring transporters for cellular uptake or efflux. Moreover, even if restricted to some enzymes and sometimes with low level activities detected, the CACO-2 presents some metabolic capacities (Scharmach et al., 2009; Borlak and Zwadlo, 2003). Some data indicated that CACO-2 metabolism can be involved in the bioactivation or the detoxification of xenobiotics including food contaminants (Fan et al., 2011; Videmann et al., 2008). Finally, some mammalian repair activities have been shown to be maintained in CACO-2 cells (Sergent et al., 2002).Therefore, all these fundamentals showed an interest in using this cell line to estimate the genotoxic potential of xenobiotics.

The CACO-2 cell line is also famous for *in vitro* pharmaco and toxicokinetics testing focused on absorption studies in which both the uptake and efflux processes and the metabolism can be followed (Yang and Yu, 2009; Fearn & Hirst, 2006). Some models of intestinal absorption have been developed from the data generated with this cell line and applied to various compounds (Prieto et al., 2010; Sun & Pang, 2008). Even though this approach is not very common with food contaminants, this topic should be of major concern considering that toxicokinetics parameters and genotoxic potential could be performed concomitantly.

USE OF CACO-2 CELLS IN GENOTOXICITY TESTS

Comet Assay

CACO-2 cells have been only scarcely used for genotoxicity testing and the major studies have been done with the alkaline comet assay detecting DNA strand breaks and alkali-labile sites. This test is known to be quick, simple and requiring only few isolated cells. Various recommendations and reviews have been published both for *in vitro* and *in vivo* studies (Burlinson et al., 2007; Hartmann et al., 2003; Tice et al., 2000). Some modifications can be applied to this assay in order to improve the specificity of the responses. One of the most common modifications is the use of Fpg, OGG1 or Endo III enzymes to increase the sensitivity of the test for detecting oxidative DNA alterations (Johansson et al., 2010; Collins, 2009).

The comet assay on CACO-2 was performed to evaluate the genotoxicity of different food contaminants. All of these experiments were done with undifferentiated cells after short term (3h) or long term (24h) exposures as recommended by Tice et al. (2000). Most of the experiments had investigated the genotoxicity of toxins (phycotoxins, cyanotoxins and mycotoxins). Some mycotoxins and their metabolites (zearalenone, deoxynivalenol and other *Fusarium* mycotoxins) were shown to induce DNA strand breaks (Abid-Essefi et al., 2009; Bony et al., 2007; Bony et al., 2006). The phycotoxin okadaic acid induced also DNA damages (Traoré et al., 2001; Valdiglesias et al., 2010). A similar response was observed with microcystin LR, a cyanotoxin produced by freshwater cyanobacteria (Zegura et al., 2008). But induction of DNA damages on CACO-2 cells was also studied with other food contaminants like the cooking process-induced compounds including heterocyclic amine PhIP and its metabolite (Vanhaecke et al., 2008), 5-hydroxymethylfurfural (Durling et al., 2009), acrylamide and its metabolite glycidamide (Puppel et al., 2005). More recently, data were published on nanoparticles (TiO2, SiO2 and ZnO) including the modified comet assay for the detection of oxidative DNA damages (Gerloff et al., 2009).

Besides food contaminants, comet assay's data on CACO-2 cells are also available in the literature on the genotoxicity or the protective effects of various compounds like drugs (Aviello et al., 2010; Kontek et al., 2010, Slamenova et al., 2006), industrial materials (Sharma et al., 2010), products of fermentation and of oxidation (Fassler et al., 2007; Duthie and Collins, 1997) or natural products contained in food (Ramos et al., 2010; Daly et al., 2009; Renis et al., 20008).

It must be noticed that other human intestinal cell lines have also been used for the comet assay as the SW480 (Naha et al., 2010), the LT97 (Knobel et al., 2006), the FHs 74 (Attene-Ramos et al., 2010), the HT29 (Kontek et al., 2010; Rosignoli et al., 2008; McCann et al., 2007; Knoll et al., 2005; Gill et al., 2005; Burns& Rowland, 2004; Ebert et al., 2001). Some data are also available, even though remaining scarce, on primary human colonocytes (Rosignoli et al., 2008; Knobel et al., 2006).

Others

A new genotoxicity assay, the γ–H2Ax assay, has recently been tested on this cell line and the effects of various polycyclic hydrocarbons have been compared (Audebert et al., 2010). Among the data, the background in CACO-2 cells was considered to be too high to generate reliable results with this assay maybe due to ineffective p53 proteins (Liu & Bodmer, 2006; Gartel et al., 1996).

Moreover, the measurement of oxidative DNA damage (8-oxoguanine) was recently conducted on the CACO-2 cell line (Ramos et al., 2010) as well as on a derived cell line exposed to iron and ascorbate (Taha et al., 2010). Some data on the detection of DNA bulky adducts in CACO-2 cells are also available in the literature (de Waard et al., 2008).

Finally, the sister chromatid exchange (SCE) was also performed on CACO-2 cells to determined the genotoxic effect of Maillard-reaction products (Glösl et al., 2004). However, the SCE assay is not commonly used for regulatory purposes nowadays.

Cytokinesis-Block Micronucleus Assay

Compared to the comet assay, the results available in the literature on the *in vitro* micronucleus assay on CACO-2 cells are rare. This assay can be performed on both differentiated and undifferentiated cells but conducting this assay on differentiated cells had required some improvements as described afterwards. For the cytokinesis-block micronucleus assay, the cell doubling time which is a key point for the cytochalasin B incubation, has been estimated around 20 h for CACO-2 cells in our culture conditions.

On Undifferentiated Cells

The micronucleus assay on undifferentiated CACO-2 cells was developed recently (Jacquin and Fessard, 2003). Till now, it has been performed in few laboratories and only to determine the mutagenic activity of natural toxins including both phycotoxins (Carvalho et al., 2006; Le Hégarat et al., 2006) and cyanotoxins (Bazin et al., 2010; Gaudin, 2007). Some of the results published in Bazin et al. (2010) on the genotoxicity of cylindrospermopsin, a freshwater cyanotoxin, are presented in the figure 1.

The HT29 human cell line has also been used to perform the micronucleus assay to study the protective effect of butyrate against oxidative stress (Hovhannisyan et al., 2009).

On Differentiated Cells: Development and Application

The only data available on the micronucleus assay on differentiated CACO-2 cells have been published by our laboratory.

One technical difficulty to perform the micronucleus assay on differentiated CACO-2 cells deals with the quiescent stage of the differentiated cells while micronucleus formation requires a cell division step. Based on the *ex vivo* micronucleus test from the spleen of treated rodents (Moore et al., 1995; Darroudi et al, 1992), we chose to treat CACO-2 cells during their differentiated stage followed by a trypsinization and a new sub-culture step. After trypsinization, cells enter into an undifferentiated stage and get through cell cycle divisions again, enabling the micronuclei to be formed.

The protocol set up took into account the recommendations of the OECD guideline n°487 for the *in vitro* micronucleus assay. After treatment (short between 3 to 6 h or long up to 24 h) of differentiated CACO-2 cells, cells are trypsinized before counting and seeding into multi-well plates. Two days after seeding, cytochalasin B (4.5 μg/ml) is added for 30 h. One hour and a half in fresh medium without cytochalasin is required to make the cells recovering from this treatment. At the end, cells are harvested by trypsinization and a slight hypotonic shock is performed to enlarge the cell cytoplasm and facilitate the scoring. For an easier visualization,the schematic representation of the schedule of this assay is presented in the figure 2.

The final built up protocol was used to study the genotoxic potential of various toxins, both cyanotoxins and phycotoxins (Bazin et al., 2010; Gaudin, 2007). To illustrate, the results published in Bazin et al. (2010) on differentiated CACO-2 cells to evaluate the genotoxic potential of cylindrospermopsin, a freshwater cyanotoxin are presented in the figure 1.

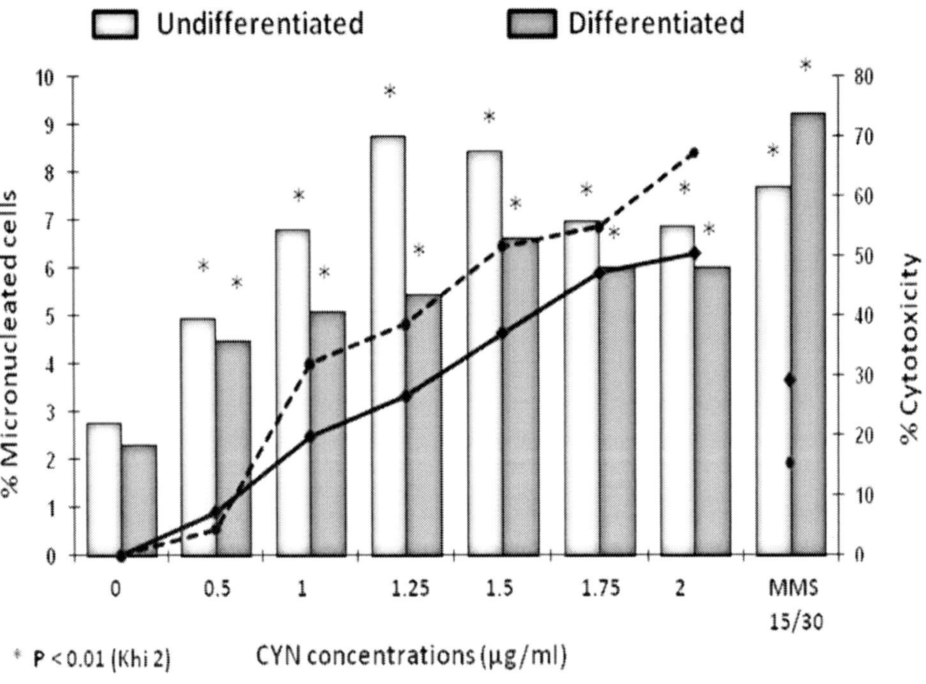

Figure 1. Micronucleus formation measured in undifferentiated and differentiated CACO-2 cells after 24h exposure to cylindrospermopsin, a freshwater cyanotoxin. The cytotoxicity was evaluated taking into account the percentage of binucleated cells in treated cultures compared to control ones (adapted from Bazin et al., 2010).

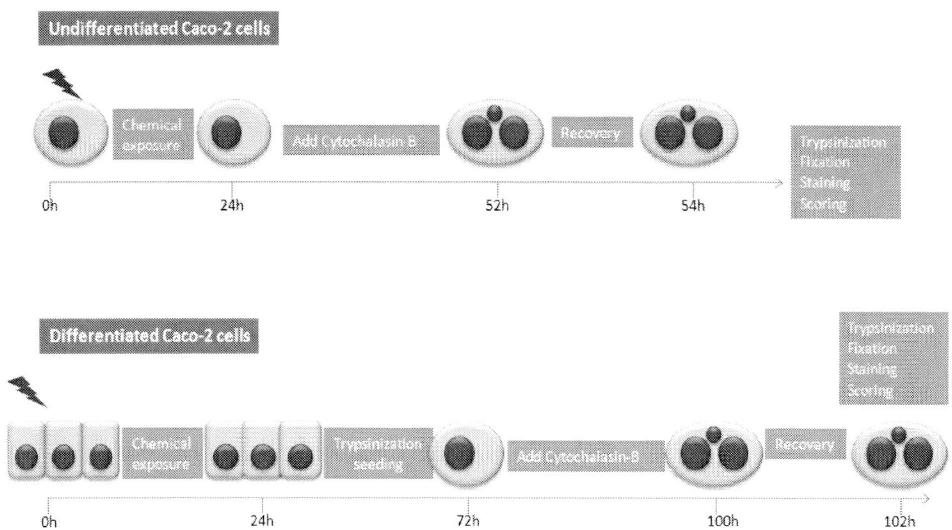

Figure 2. Schedule of the cytokinesis-block micronucleus assay on undifferentiated (1 day of subculture) and differentiated (21 days after confluence) CACO-2 cells.

Improvement of this Cell Model to Predict Toxicity and Mechanistic Endpoints

More complex cell systems can also been developed to obtain integrative results. For example, the CACO-2 cells can be cultured on inserts delimiting two compartments, one apical and one basal. In this case, another cell type can be added to the lower compartment (Kure et al., 2010; Kadiyala et al., 2010). Another possibility is to prepare some co-cultures whether with macrophages and dendritic cells (Leonard et al., 2010), with myofibroblasts (Seltana et al., 2010) or with mucus-secreting like cells (HT29-MTX) (Mahler et al., 2009; Hilgendord et al., 2000; Walter et al., 1996) or endothelial cells (Schanz et al., 2010). Some of these complex systems have been used for cytotoxicity or toxicokinetics studies but are promising also for genotoxicity investigation.

Co-Cultures CACO-2/TK6

As recommended by most experts, the predictivity of mutagenicity tests should be improved by developing more accurate *in vitro* models (Kirkland et al., 2007). Conventional single culture models commonly used for *in vitro* genotoxicity assays might not be enough relevant to evaluate human mutagenicity, generating both false-positive and false-negative results. Therefore, new testing methods for safety and toxicity without requiring the use of animals should be developed. The intestinal first-pass effect involving both metabolic enzymes and transporters, including CYP450 and MDRs, is considered to regulate bioavailability, blood concentration and systemic toxicity of xenobiotics including food contaminants. Developing CACO-2 coculture models should be based on the complexity of these metabolic processes occurring *in vivo* in the intestine. Therefore, a co-culture of CACO-

2 cells monolayer and human lymphoblastoid TK6 cells has been proposed to improve the *in vitro* intestinal model. In this case, differentiated CACO-2 cells mimic the intestinal compartment and the TK6 cells the blood compartment. In this system, the final toxicity on TK6 cells could be influenced by the CACO-2 cells kinetics. We investigated the genotoxicity of food contaminants like benzo[a]pyrene (B[a]P) and aflatoxin B1 (AFB1) in this co-culture CACO-2/TK6 model. CACO-2 cells were seeded on inserts and left to differentiate before treatment for 24 h or 48h by apical addition of the compound. Prior to treatment, dividing TK6 cells were added into the basal compartment. After treatment, the comet assay was performed on both cell lines as well as the micronucleus assay on the TK6 cells only. The positive control Methylmethanesulfonate MMS (10 μg/ml) induced genotoxic damage in both cell lines, including micronucleus formation in TK6 cells. B[a]P also induced comet formation in a dose dependent-manner between 5 and 50 μM at 24h. However, AFB1 did not induce comet formation in this system (unpublished results). Our results suggest that the co-culture CACO-2/TK6 could be an attractive model for genetic toxicology taking into account intestinal absorption and metabolism.

Metabolism Regulation

Considering that CYP3A represents the main drug-metabolizing system in the small intestine, accounting for approximately 80% of the total CYP450 content (Kolars et al., 1994), the *in vitro* models used to mimic the intestinal compartment need to express relevant phase I and phase II enzymes. It has been demonstrated that some clones of CACO-2 cells failed to express substantial amounts of CYP3A4 (Schmiedlin-Ren et al., 1997). Considering that the majority of mutagens require metabolic activation to become genotoxic, the capacities of CACO-2 cells in monolayers or in co-cultures can be improved to increase their relevance for genetic toxicology studies. As an example, the amounts of CYP3A4 expressed in CACO-2 cells can be increased by incubating the cells with an inducer like the 1α,25-dihydroxyvitamin D3, the biological active form of vitamin D and the natural ligand of the Vitamin D receptor (Thummel et al., 2001). The authors demonstrated that CACO-2 cells monolayers with induced CYP34A are closer, in term of first-pass intestine metabolism, to the duodenal and jejunal cells than the parental CACO-2 cells.

Inhibition of CYP450 enzymes can also be used to demonstrate the involvement if this enzymatic system in the genotoxicity of food contaminants. The micronuclei formation in CACO-2 cells induced by the cyanotoxin cylindrospermopsin and the positive control aflatoxin B1 was shown to be reduced when the cells were previously treated with ketoconazole, an inhibitor of CYP450 (1A and 3A4) (Bazin et al., 2010) as shown in figure 3. This can help in investigating the mechanism of genotoxicity and defining which metabolic enzymes are implicated in the effect. Some other cell lines obtained by CACO-2 cells transfection can be useful to study the influence of transporters like MDR1 (Hilgendorf et al., 2005) and their involvement in the modulation of the genotoxicity of xenobiotics.

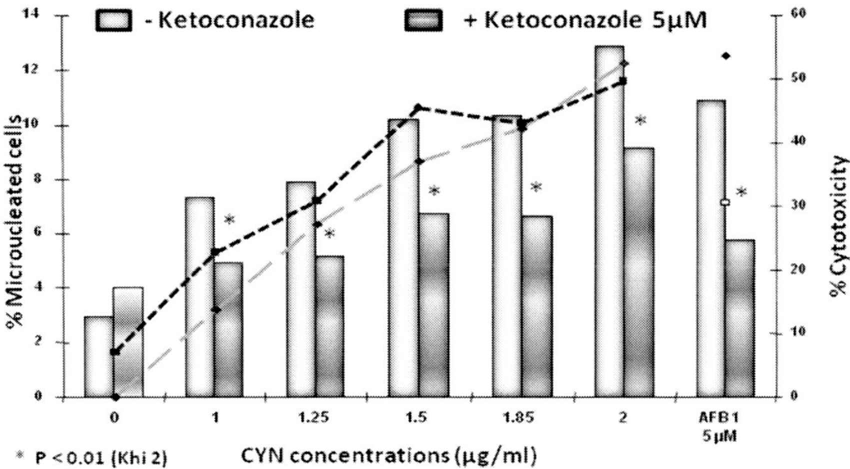

Figure 3. Micronucleus formation measured in undifferentiated CACO-2 cells after 24h exposure to cylindrospermopsin, a freshwater cyanotoxin. Ketoconazole was added 2h prior to cylindrospermopsin exposure. The cytotoxicity was evaluated taking into account the percentage of binucleated cells in treated cultures compared to control ones (adapted from Bazin et al., 2010).

Conclusions

In general, the CACO-2 cell line could be considered as a good predictive model in genetic toxicology compared to other available cell lines expressing poor characteristics of human metabolism and transporters. This model remains one of the most studied intestinal cell line and one of the most representative for enterocytes. Moreover, numerous data are available in the literature for this cell line on expression of genes and proteins involved in metabolism, on permeability studies of various xenobiotics as well as on comparisons with human intestine. One argument against CACO-2 cells to be used in genotoxicity testing is the absence of functional p53 due to a mutation in the codon 206. However, efforts have to be made to increase the database on the genotoxicity results obtained on this cell line with different kind of compounds in order to estimate its sensitivity and its specificity in genotoxicity testing, implementing its relevance for the study of food contaminants.

Acknowledgment

We thank Annick MOUROT, Sylvie HUET and Emmanuelle BAZIN for their technical assistance in the genotoxicity testing.

References

Abid-Essefi, S., Bouaziz, C., Golli-Bennour, E.E., Ouanes, Z., Bacha, H. (2009). Comparative study of toxic effects of zearalenone and its two major metabolites alphazearalenol and beta-zearalenol on cultured human CACO-2 cells. *J Biochem Mol Toxicol.,* 23(4), 233-243.

Anderle, P., Rakhmanova, V., Woodford, K., Zerangue, N., Sadée, W. (2003). Messenger RNA expression of transporter and ion channel genes in undifferentiated and differentiated CACO-2 cells compared to human intestines. *Pharm Res*., 20(1), 3-15.

Attene-Ramos, M.S., Nava, G.M., Muellner, M.G., Wagner, E.D., Plewa, M.J., Gaskins, H.R. (2010). DNA damage and toxicogenomic analyses of hydrogen sulfide in human intestinal epithelial FHs 74 Int cells. *Environ Mol Mutagen*., 51(4), 304-314.

Audebert, M., Riu, A., Jacques, C., Hillenweck, A., Jamin, E.L., Zalko, D., Cravedi, J.P. (2010). Use of the γH2AX assay for assessing the genotoxicity of polycyclic aromatic hydrocarbons in human cell lines. *Toxicol Lett.* Sep 9.

Aviello, G., Rowland, I., Gill, C.I., Acquaviva, A.M., Capasso, F., McCann, M., Capasso, R., Izzo A.A., Borrelli, F. (2010) Anti-proliferative effect of rhein, an anthraquinone isolated from Cassia species, on CACO-2 human adenocarcinoma cells. *Journal of Cellular and Molecular Medicine* 14 (7), pp. 2006-2014.

Bazin, E., Mourot, A., Humpage, A.R., Fessard, V. (2010). Genotoxicity of a freshwater cyanotoxin, cylindrospermopsin, in two human cell lines: CACO-2 and HepaRG. *Environ Mol Mutagen*., 51(3), 251-259.

Bony, S., Carcelen, M., Olivier, L., Devaux, A. (2006). Genotoxicity assessment of deoxynivalenol in the CACO-2 cell line model using the Comet assay. *Toxicol Lett*., 166(1), 67-76.

Bony, S., Olivier-Loiseau, L., Carcelen, M., Devaux, A. (2007). Genotoxic potential associated with low levels of the *Fusarium* mycotoxins nivalenol and fusarenon X in a human intestinal cell line. *Toxicol In Vitro*., 21(3),457-465.

Borlak, J. & Zwadlo, C. (2003) Expression of drug-metabolizing enzymes, nuclear transcription factors and ABC transporters in CACO-2 cells. *Xenobiotica* 33(9):927-943.

Brambilla, G., Mattioli, F., Robbiano, L., Martelli, A. (2010). Genotoxicity and carcinogenicity testing of pharmaceuticals: correlations between induction of DNA lesions and carcinogenic activity. *Mutat Res.,* 705(1), 20-39.

Brandon, E.F., Bosch, T.M., Deenen, M.J., Levink, R., van der Wal, E., van Meerveld, J.B., Bijl, M., Beijnen, J.H., Schellens, J.H., Meijerman, I. (2006). Validation of *in vitro* cell models used in drug metabolism and transport studies; genotyping of cytochrome P450, phase II enzymes and drug transporter polymorphisms in the human hepatoma (HepG2), ovarian carcinoma (IGROV-1) and colon carcinoma (CACO-2, LS180) cell lines. *Toxicol. Appl. Pharmacol*., 211(1), 1-10.

Burlinson, B., Tice, R.R., Speit, G., Agurell, E., Brendler-Schwaab, S.Y., Collins, A.R., Escobar, P., Honma, M., Kumaravel, T.S., Nakajima, M., Sasaki, Y.F., Thybaud, V., Uno, Y., Vasquez, M., Hartmann, A. (2007). *In vivo* comet assay Workgroup, part of the Fourth International Workgroup on Genotoxicity Testing. Fourth International Workgroup on Genotoxicity testing: results of the *in vivo* Comet assay workgroup. *Mutat Res*., 627(1), 31-35.

Burns, A.J. and Rowland, I.R. (2004). Antigenotoxicity of probiotics and prebiotics on faecal water-induced DNA damage in human colon adenocarcinoma cells. *Mutat Res*., 551(1-2), 233-243.

Calcagno, A.M., Ludwig, J.A., Fostel, J.M., Gottesman, M.M., Ambudkar, S.V. (2006). Comparison of drug transporter levels in normal colon, colon cancer, and CACO-2 cells: impact on drug disposition and discovery. *Molecular Pharmaceutics*, 3 (1), 87-93.

Carvalho, P.S., Catian, R., Moukha, S., Matias, W.G., Creppy, E.E. (2006). Comparative study of Domoic Acid and Okadaic Acid induced-chromosomal abnormalities in the CACO-2 cell line. *Int J Environ Res Public Health.*, 3(1), 4-10.

Chantret, I., Barbat, A., Dussaulx, E., Brattain, M.G., Zweibaum, A. (1988). Epithelial polarity, villin expression, and enterocytic differentiation of cultured human colon carcinoma cells: a survey of twenty cell lines. *Cancer Res.*, 48(7), 1936-1942.

Cimino, M.C. (2006). Comparative overview of current international strategies and guidelines for genetic toxicology testing for regulatory purposes. *Environ Mol Mutagen*, 47, 362-390.

Coffing, S., Engel, M., Dickinson, D., Thiffeault, C., Spellmann, R., Shutsky, T., Schuler, M. (2010). The rat gut micronucleus assay: A good choice for alternative *in vivo* genetic toxicology testing strategies. *Environ Mol Mutagen.* Sep 24.

Collins, A.R. (2009). Investigating oxidative DNA damage and its repair using the comet assay. *Mutat Res.*, 681(1), 24-32.

Daly, T.J., Aherne, S.A., O'Connor, T.P., O'Brien, N.M. (2009). Lack of genoprotective effect of phytosterols and conjugated linoleic acids on CACO-2 cells. *Food Chem Toxicol.,* 47(8), 1791-1796.

Darroudi, F., Farooqi, Z., Benova, D., Natarajan, A.T. (1992). The mouse splenocyte assay, an *in vivo/in vitro* system for biological monitoring: studies with X-rays, fission neutrons and bleomycin. *Mutat Res.*, 272(3), 237-248.

Durling, L.J., Busk, L., Hellman, B.E. (2009). Evaluation of the DNA damaging effect of the heat-induced food toxicant 5-hydroxymethylfurfural (HMF) in various cell lines with different activities of sulfotransferases. *Food Chem Toxicol.*, 47(4), 880-884.

Duthie, S.J. & Collins, A.R. (1997). The influence of cell growth, detoxifying enzymes and DNA repair on hydrogen peroxide-mediated DNA damage (measured using the comet assay) in human cells. *Free Radic Biol Med.,* 22(4), 717-724.

Ebert, M.N., Beyer-Sehlmeyer, G., Liegibel, U.M., Kautenburger, T., Becker, T.W., Pool-Zobel, B.L. (2001). Butyrate induces glutathione S-transferase in human colon cells and protects from genetic damage by 4-hydroxy-2-nonenal. *Nutr Cancer.*, 41(1-2), 156-164.

EC 1223/2009 Regulation of the European Parliament and of the Council of 30 November 2009 on cosmetic products Text with EEA relevance. *Official Journal* L 342 , 22/12/2009 P. 0059 – 0209

http://eur-lex.europa.eu/LexUriServ/LexUriServ.do?uri=OJ:L:2009:342:0059:01:EN:HTML

Englund, G., Rorsman, F., Rönnblom, A., Karlbom, U., Lazorova, L., Gråsjö, J., Kindmark, A., Artursson, P. (2006). Regional levels of drug transporters along the human intestinal tract: co-expression of ABC and SLC transporters and comparison with CACO-2 cells. Eur J Pharm Sci., 29(3-4), 269-277.

Engman, H.A., Lennernas, H., Taipalensuu, J., Otter, C., Leidvik, B., Artursson, P. (2001). CYP3A4, CYP3A5, and MDR1 in human small and large intestinal cell lines suitable for drug transport studies. *J Pharm Sci.,* 90(11), 1736-1751.

Fan, J., Maeng, H.J., Du, Y., Kwan, D., Pang, K.S. (2011). Transport of 5,5-diphenylbarbituric acid and its precursors and their effect on P-gp, MRP2 and CYP3A4 in CACO-2 and LS180 cells. *Eur J Pharm Sci.,* 42 (1-2), 19-29.

Fässler, C., Gill, C.I., Arrigoni, E., Rowland, I., Amadò, R,. (2007). Fermentation of resistant starches: influence of *in vitro* models on colon carcinogenesis. *Nutr Cancer.*, 58(1), 85-92.

Fearn, R.A. & Hirst, B.H. (2006). Predicting oral drug absorption and hepatobiliary clearance: Human intestinal and hepatic *in vitro* cell models. *Environmental Toxicology and PharmacologyVolume 21*, Issue 2, February, Pages 168-178.

Fenech, M. (2000). The *in vitro* micronucleus technique. *Mutat Res.*, 455(1-2), 81-95.

Fenech, M. (2007). Cytokinesis-block micronucleus cytome assay. *Nat Protoc.*, 2(5):1084-1104.

Gartel, A.L, Serfas, M.S., Gartel, M., Goufman, E., Wu, G.S., el-Deiry, W.S., Tyner, A.L. (1996). p21 (WAF1/CIP1) expression is induced in newly nondividing cells in diverse epithelia and during differentiation of the CACO-2 intestinal cell line. *Exp Cell Res.*, 227(2), 171-181.

Gaudin, J. (2007). Contribution à l'évluation de la génotoxicité par des tests in vitro et in vivo : application à des toxins de cyanobactéries, les microcystines –LR et –RR. PhD thesis, university of Rennes 1

Gerloff, K., Albrecht, C., Boots, A.W., Frster, I., Schins, R.P.F. (2009). Cytotoxicity and oxidative DNA damage by nanoparticles in human intestinal CACO-2 cells. *Nanotoxicology*, 3 (4), 355-364.

Gill, C.I.R., Boyd, A., McDermott, E., McCann, M., Servili, M., Selvaggini, R., Taticchi, A., Esposto S., Montedoro J., McGlynn H., Rowland, I. (2005). Potential anti-cancer effects of virgin olive oil phenols on colorectal carcinogenesis models *in vitro. International Journal of Cancer*, 117 (1), 1-7.

Glösl, S., Wagner, K.H., Draxler, A., Kaniak, M., Lichtenecker, S., Sonnleitner, A., Somoza, V., Erbersdobler, H., Elmadfa, I. (2004). Genotoxicity and mutagenicity of melanoidins isolated from a roasted glucose-glycine model in human lymphocyte cultures, intestinal CACO-2 cells and in the Salmonella typhimurium strains TA98 and TA102 applying the AMES test. *Food Chem Toxicol.*, 42(9),1487-1495.

Hartmann, A., Agurell, E., Beevers, C., Brendler-Schwaab, S., Burlinson, B., Clay, P., Collins, A., Smith, A., Speit, G., Thybaud, V., Tice, R.R. (2003). Recommendations for conducting the *in vivo* alkaline Comet assay. 4th International Comet Assay Workshop. *Mutagenesis*, 18(1), 45-51.

Hayeshi, R., Hilgendorf, C., Artursson, P., Augustijns, P., Brodin, B., Dehertogh, P., Fisher, K., Fossati, L., Hovenkamp, E., Korjamo, T., Masungi, C., Maubon, N., Mols, R., Müllertz, A., Mönkkönen, J., O'Driscoll, C., Oppers-Tiemissen, H.M., Ragnarsson, E.G., Rooseboom, M., Ungell, A.L. (2008). Comparison of drug transporter gene expression and functionality in CACO-2 cells from 10 different laboratories. *Eur J Pharm Sci.*, 35(5), 383-396.

Hilgendorf, C., Ahlin, G., Seithel, A., Artursson, P., Ungell, A.L., Karlsson, J. (2007). Expression of thirty-six drug transporter genes in human intestine, liver, kidney, and organotypic cell lines. *Drug Metab Dispos.*, 35(8), 1333-1340.

Hilgendorf, C., Spahn-Langguth, H., Regårdh, C.G., Lipka, E., Amidon, G.L., Langguth, P. (2000). CACO-2 versus CACO-2/HT29-MTX co-cultured cell lines: permeabilities via diffusion, inside- and outside-directed carrier-mediated transport. *J Pharm Sci.*, 89(1), 63-75.

Hilgendorf, C., Spahn-Langguth, H., Rhedin, M., Regårdh, C.G., Löwenadler, B., Langguth, P. (2005). Selective downregulation of the MDR1 gene product in CACO-2 cells by stable transfection to prove its relevance in secretory drug transport. *Mol Pharm.*, 2(1), 64-73.

Hovhannisyan, G., Aroutiounian, R., Glei, M. (2009). Butyrate reduces the frequency of micronuclei in human colon carcinoma cells *in vitro*. *Toxicol In Vitro.,* 23(6), 1028-1033.

Jacquin, A.G. & Fessard, V. (2003). Development of the *in vitro* micronucleus assay on an intestinal model, the CACO-2 cell line. *European Environmental Mutagen Society,* Aberdeen (UK)

Johansson, C., Møller, P., Forchhammer, L., Loft, S., Godschalk, R.W., Langie, S.A., Lumeij, S., Jones, G.D., Kwok, R.W., Azqueta, A., Phillips, D.H., Sozeri, O., Routledge, M.N., Charlton, A.J., Riso, P., Porrini, M., Allione, A., Matullo, G., Palus, J., Stepnik, M., Collins, A.R., Möller, L. (2010). An ECVAG trial on assessment of oxidative damage to DNA measured by the comet assay. *Mutagenesis*, 25(2), 125-132.

Kadiyala, I., Loo, Y., Roy, K., Rice, J., Leong, K.W. (2010). Transport of chitosan-DNA nanoparticles in human intestinal M-cell model versus normal intestinal enterocytes. *Eur J Pharm Sci.*, 39(1-3), 103-109.

Kirkland, D., Pfuhler, S., Tweats, D., Aardema, M., Corvi, R., Darroudi, F., Elhajouji, A., Glatt, H., Hastwell, P., Hayashi, M., Kasper, P., Kirchner, S., Lynch, A., Marzin, D., Maurici, D., Meunier, J.R., Müller, L., Nohynek, G., Parry, J., Parry, E., Thybaud, V., Tice, R., van Benthem, J., Vanparys, P., White, P. (2007). How to reduce false positive results when undertaking in vitro genotoxicity testing and thus avoid unnecessary follow-up animal tests: Report of an ECVAM Workshop. *Mutat Res.,* 628(1),31-55.

Kirsch-Volders, M., Elhajouji, A., Cundari, E., Van Hummelen, P. (1997). The *in vitro* micronucleus test: a multi-endpoint assay to detect simultaneously mitotic delay, apoptosis, chromosome breakage, chromosome loss and non-disjunction. *Mutat Res.,* 392(1-2), 19-30.

Kirsch-Volders, M., Sofuni, T., Aardema, M., Albertini, S., Eastmond, D., Fenech, M., Ishidate, M., Kirchner, S., Lorge, E., Morita, T., Norppa, H., Surralles, J., Vanhauwaert, A., Wakata, A. (2003). Report from the *in vitro* micronucleus assay working group. *Mutat Res,* 540, 153-163.

Knöbel, Y., Glei, M., Weise, A., Osswald, K., Schäferhenrich, A., Richter, K.K., Claussen, U., Pool-Zobel, B.L. (2006). Uranyl nitrilotriacetate, a stabilized salt of uranium, is genotoxic in nontransformed human colon cells and in the human colon adenoma cell line LT97. *Toxicol Sci.,* 93(2), 286-297.

Knoll, N., Ruhe, C., Veeriah, S., Sauer, J., Glei, M., Gallagher, E.P., Pool-Zobel, B.L. (2005). Genotoxicity of 4-hydroxy-2-nonenal in human colon tumor cells is associated with cellular levels of glutathione and the modulation of glutathione S-transferase A4 expression by butyrate. *Toxicol Sci.,* 86(1), 27-35.

Kolars, J.C., Lown, K.S., Schmiedlin-Ren, P., Ghosh, M., Fang, C., Wrighton, S.A., Merion, R.M., Watkins, P.B. (1994). CYP3A gene expression in human gut epithelium. *Pharmacogenetics*, 4(5), 247-259.

Kontek, R., Matlawska-Wasowska, K., Kalinowska-Lis, U., Marciniak, B. (2010). Genotoxic effects of irinotecan combined with the novel platinum(II) complexes in human cancer cells. *Chemico-Biological Interactions*, 188, 66-74.

Kure, I., Nishiumi, S., Nishitani, Y., Tanoue, T., Ishida, T., Mizuno, M., Fujita, T., Kutsumi, H., Arita, M., Azuma, T., Yoshida, M. (2010). Lipoxin A(4) reduces lipopolysaccharide-induced inflammation in macrophages and intestinal epithelial cells through inhibition of nuclear factor-kappaB activation. *J Pharmacol Exp Ther.*, 332(2), 541-548.

Le Hégarat, L., Jacquin, A.G., Bazin, E., Fessard, V. (2006). Genotoxicity of the marine toxin okadaic acid, in human CACO-2 cells and in mice gut cells. *Environ Toxicol.*, 21(1), 55-64.

Leonard, F., Collnot, E.M., Lehr, C.M. (2010). A Three-Dimensional Coculture of Enterocytes, Monocytes and Dendritic Cells To Model Inflamed Intestinal Mucosa *in Vitro*. *Mol Pharm.*

Liu, Y. & Bodmer, W.F. (2006). Analysis of P53 mutations and their expression in 56 colorectal cancer cell lines. *Proc Natl Acad Sci U S A.,* 103(4), 976-981.

Mahler, G.J., Shuler, M.L., Glahn, R.P. (2009). Characterization of CACO-2 and HT29-MTX cocultures in an *in vitro* digestion/cell culture model used to predict iron bioavailability. *J Nutr Biochem.*, 20(7), 494-502.

Maubon, N., Le Vee, M., Fossati, L., Audry, M., Le Ferrec, E., Bolze , S., Fardel, O. (2007). Analysis of drug transporter expression in human intestinal CACO-2 cells by real-time PCR. *Fundam Clin Pharmacol.*, 21(6), 659-663.

McCann, M.J., Gill, C.I., O' Brien, G., Rao, J.R., McRoberts, W.C., Hughes, P., McEntee, R., Rowland, I.R. (2007). Anti-cancer properties of phenolics from apple waste on colon carcinogenesis *in vitro*. *Food Chem Toxicol.*, 45(7), 1224-1230.

Moore, F.R., Urda, G.A., Krishna, G., Theiss, J.C. (1995). An *in vivo/in vitro* method for assessing micronucleus and chromosome aberration induction in rat bone marrow and spleen. 1. Studies with cyclophosphamide. *Mutat Res.*, 335(2), 191-199.

Naha, P.C., Bhattacharya, K., Tenuta, T., Dawson, K.A., Lynch, I., Gracia, A., Lyng, F.M., Byrne, H.J. (2010). Intracellular localisation, geno- and cytotoxic response of polyN-isopropylacrylamide (PNIPAM) nanoparticles to human keratinocyte (HaCaT) and colon cells (SW 480). *Toxicol Lett.,* 198(2), 134-143.

OECD Guidelines for the Testing of Chemicals Section 4: Health Effects http://oberon.sourceoecd.org/vl=430571/cl=15/nw=1/rpsv/cw/vhosts/oecdjournals/16073 10x/v1n4/contp1-1.htm
http://www.oecd.org/document/55/0,3343,en_2649_34377_2349687_1_1_1_1,00.html

Parry, J. (2000). Guidance on a strategy for testing of chemicals for mutagenicity. *"Committee on mutagenicity of chemicals in food, consumer products and the environment"*

Prieto, P., Hoffmann, S., Tirelli, V., Tancredi, F., González, I., Bermejo, M., De Angelis, I. (2010). An exploratory study of two CACO-2 cell models for oral absorption: A report on their within-laboratory and between-laboratory variability, and their predictive capacity. *Altern Lab Anim.*, 38(5), 367-386.

Puppel, N., Tjaden, Z., Fueller, F., Marko, D. (2005). DNA strand breaking capacity of acrylamide and glycidamide in mammalian cells. *Mutat Res.,* 580(1-2), 71-80.

Ramos, A.A., Pereira-Wilson, C., Collins, A.R. (2010). Protective effects of ursolic acid and luteolin against oxidative DNA damage include enhancement of DNA repair in CACO-2 cells. *Mutat Res.*, 692(1-2), 6-11.

Renis, M., Calandra, L., Scifo, C., Tomasello, B., Cardile, V., Vanella, L., Bei, R., La Fauci, L., Galvano, F. (2008). Response of cell cycle/stress-related protein expression and DNA damage upon treatment of CACO-2 cells with anthocyanins. *Br J Nutr.*, 100(1), 27-35.

Rosignoli, P., Fabiani, R., De Bartolomeo, A., Fuccelli, R., Pelli, M.A., Morozzi, G. (2008). Genotoxic effect of bile acids on human normal and tumour colon cells and protection by dietary antioxidants and butyrate. *Eur J Nutr.*, 47(6), 301-309.

Sasaki, Y.F., Sekihashi, K., Izumiyama, F., Nishidate, E., Saga, A., Ishida, K., Tsuda, S. (2000). The comet assay with multiple mouse organs: comparison of comet assay results and carcinogenicity with 208 chemicals selected from the IARC monographs and U.S. NTP Carcinogenicity Database. *Crit Rev Toxicol.*, 30(6), 629-799.

Schanz, J., Pusch, J., Hansmann, J., Walles, H. (2010). Vascularised human tissue models: a new approach for the refinement of biomedical research. *J Biotechnol.*, 148(1), 56-63.

Scharmach, E., Hessel, S., Niemann, B., Lampen, A. (2009). Glutathione S-transferase expression and isoenzyme composition during cell differentiation of CACO-2 cells. *Toxicology*, 265(3), 122-126.

Schmiedlin-Ren, P., Thummel, K.E., Fisher, J.M., Paine, M.F., Lown, K.S., Watkins, P.B. (1997). Expression of enzymatically active CYP3A4 by CACO-2 cells grown on extracellular matrix-coated permeable supports in the presence of 1alpha,25-dihydroxyvitamin D3. *Mol Pharmacol.*, 51(5), 741-754.

Seithel, A., Karlsson, J., Hilgendorf, C., Björquist, A., Ungell, A.L. (2006). Variability in mRNA expression of ABC- and SLC-transporters in human intestinal cells: comparison between human segments and CACO-2 cells. *Eur J Pharm Sci.*, 28(4), 291-299.

Seltana, A., Basora, N., Beaulieu, J.F. (2010). Intestinal epithelial wound healing assay in an epithelial-mesenchymal co-culture system. *Wound Repair Regen.*, 18(1), 114-122.

Sergent, C., Franco, N., Chapusot, C., Lizard-Nacol, S., Isambert, N., Correia, M., Chauffert, B. (2002). Human colon cancer cells surviving high doses of cisplatin or oxaliplatin *in vitro* are not defective in DNA mismatch repair proteins. *Cancer Chemother Pharmacol.*, 49(6), 445-452.

Sharma, A.K., Schmidt, B., Frandsen, H., Jacobsen, N.R., Larsen, E.H., Binderup, M.L. (2010). Genotoxicity of unmodified and organo-modified montmorillonite. *Mutat Res.*, 700(1-2), 18-25.

Slamenová, D., Horváthová, E., Bartková, M. (2006). Nature of DNA lesions induced in human hepatoma cells, human colonic cells and human embryonic lung fibroblasts by the antiretroviral drug 3'-azido-3'-deoxythymidine. *Mutat Res.*, 593(1-2), 97-107.

Sun, H. & Pang, K.S. (2008). Permeability, transport, and metabolism of solutes in CACO-2 cell monolayers: a theoretical study. *Drug Metab Dispos.*, 36(1), 102-123.

Taha, R., Seidman, E., Mailhot, G., Boudreau, F., Gendron, F.P., Beaulieu, J.F., Ménard, D., Delvin, E., Amre, D., Levy, E. (2010). Oxidative stress and mitochondrial functions in the intestinal CACO-2/15 cell line. *PLoS One.*, 5(7), e11817.

Taipalensuu, J., Törnblom, H., Lindberg, G., Einarsson, C., Sjöqvist, F., Melhus, H., Garberg, P., Sjöström, B., Lundgren, B., Artursson, P. (2001). Correlation of gene expression of ten drug efflux proteins of the ATP-binding cassette transporter family in normal human jejunum and in human intestinal epithelial CACO-2 cell monolayers. *J Pharmacol Exp Ther.*, 299(1), 164-170.

Thummel, K.E., Brimer, C., Yasuda, K., Thottassery, J., Senn, T., Lin, Y., Ishizuka, H., Kharasch, E., Schuetz, J., Schuetz, E. (2001). Transcriptional control of intestinal cytochrome P-4503A by 1alpha,25-dihydroxy vitamin D3. *Mol Pharmacol.*, (60), 1399-1406.

Tice RR, Agurell E, Anderson D, Burlinson B, Hartmann A, Kobayashi H, Miyamae Y, Rojas E, Ryu JC, Sasaki YF. Single cell gel/comet assay: guidelines for in vitro and in vivo genetic toxicology testing. *Environ Mol Mutagen.* 2000;35(3):206-21.

Traoré, A., Baudrimont, I., Ambaliou, S., Dano, S.D., Creppy, E.E. (2001). DNA breaks and cell cycle arrest induced by okadaic acid in CACO-2 cells, a human colonic epithelial cell line. *Arch Toxicol.*, 75(2), 110-117.

Valdiglesias, V., Méndez, J., Pásaro, E., Cemeli, E., Anderson, D., Laffon, B. (2010). Assessment of okadaic acid effects on cytotoxicity, DNA damage and DNA repair in human cells. *Mutat Res.*, 689(1-2), 74-79.

Vanhaecke, L., Derycke, L., Le Curieux, F., Lust, S., Marzin, D., Verstraete, W., Bracke, M. (2008). The microbial PhIP metabolite 7-hydroxy-5-methyl-3-phenyl-6,7,8,9-tetrahydropyrido[3′,2′:4 ,5]imidazo[1,2-a]pyrimidin-5-ium chloride (PhIP-M1) induces DNA damage, apoptosis and cell cycle arrest towards CACO-2 cells. *Toxicology Letters,* 178 (1), 61-69.

Vanhauwaert, A., Vanparys, P., Kirsch-Volders, M. (2001).The *in vivo* gut micronucleus test detects clastogens and aneugens given by gavage. *Mutagenesis*, 16(1), 39-50.

Videmann, B., Mazallon, M., Tep, J., Lecoeur, S. (2008). Metabolism and transfer of the mycotoxin zearalenone in human intestinal CACO-2 cells. *Food Chem Toxicol.*, 46(10), 3279-3286.

Walter, E., Janich, S., Roessler, B.J., Hilfinger, J.M., Amidon, G.L. (1996). HT29-MTX/CACO-2 cocultures as an *in vitro* model for the intestinal epithelium: *in vitro-in vivo* correlation with permeability data from rats and humans. *J Pharm Sci.*, 85(10), 1070-1076.

de Waard, P.W., de Kok, T.M., Maas, L.M., Peijnenburg, A.A., Hoogenboom, R.L., Aarts, J.M., van Schooten, F.J. (2008). Influence of TCDD and natural Ah receptor agonists on benzo[a]pyrene-DNA adduct formation in the CACO-2 human colon cell line. *Mutagenesis*, 23(1), 67-73.

Yang Y. and Yu L.X. Oral Drug Absorption, Evaluation, and Prediction. In *"Developing Solid Oral Dosage Forms"*, 2009, Pages 289-308.

Zegura, B., Volčič, M., Lah, T.T., Filipič, M. (2008). Different sensitivities of human colon adenocarcinoma (CACO-2), astrocytoma (IPDDC-A2) and lymphoblastoid (NCNC) cell lines to microcystin-LR induced reactive oxygen species and DNA damage. *Toxicon*, 52 (3), 518-525.

In: CACO-2 Cells and their Uses
Editor: Megan A. Schulz, pp. 129-163
ISBN 978-1-61209-917-0

Chapter 9

CACO-2 CELLS FOR STUDYING INTESTINAL ABSORPTION MECHANISMS: COMPARISON AND CORRELATIONS WITH RAT MODEL

Victor Mangas-Sanjuan, Marta Gonzalez-Alvarez, Isabel Gonzalez-Alvarez and Marival Bermejo Sanz
Universidad Miguel Hernández, Spain

ABSTRACT

CACO-2 cell model has evolved as a useful tool for studying human oral drug absorption. The model is suitable for a highthrougput system, in order to rank compounds in the screening phase, but it could be also used for mechanistic studies by adapting the experimental conditions, for instance using different initial concentrations and performing bidirectional experiments. With the adequate experimental design and data treatment it allows to explore the permeation mechanism (i.e passive versus carrier mediated) and to evaluate the transport parameters. In this chapter the use of CACO-2 cells for prediction human oral fraction absorbed will be discussed as well as the advantages and disadvantages versus other in vitro models (as PAMPA membranes) or in situ models as the rat perfusion method. In second place, the different mathematical approaches to obtain transport parameters from the CACO-2 model will be described and discussed in terms of experimental burden, accuracy and relevance for the aim of the study. Finally several examples of mechanistic studies of fluoroquinolones will be used to illustrate the detection of secretion mechanism and the correlation with the results obtained in rat. Examples will be provided from our experimental experience on the study of fluoroquinolones but results from other papers will be also reviewed and discussed. This short review does not intend to be a comprehensive compilation of all the CACO-2 related information and published experiments but to point out some useful examples in order to help the users to optimize the use of this model.

INTRODUCTION: IN VITRO MODELS OF INTESTINAL PERMEATION

As oral route of administration is the most used and convenient for patients, it is important to characterize the absorbability of drug candidates at an early stage of their development. Consequently, studies on drug intestinal absorption have received considerable attention. Animal models have been used to perform absorption studies but clearly these models are not suiTable as a high throughput screening tool. This fact led to the development of in vitro methods of the intestinal membrane. As reviewed in some papers (Balimane et al., 2000,Nigsch et al., 2007) a number of in vitro models for intestinal absorption are available, which can be classified in physicochemical and biological methods. The advantages of integrated approaches, by using a combination of models have been presented (Miret et al., 2004). However, the cell based methods alone are currently the most widely used for screening purposes, and among them the CACO-2 cell line has been extensively studied (Artursson, 1991,Artursson et al., 2001,Rubas et al., 1996,Yamashita et al., 2000). These cells are able to differentiate in long-term culture and to polarize, when seeded on semi-permeable membranes, on which they form a continuous monolayer with tight junctions, mimicking the intestinal barrier. Drug transport studies in epithelial cell monolayers grown on permeable supports are easy to perform under controlled conditions and the model structure permits bidirectional studies from apical to basolateral and from basolateral to apical from which relevant mechanistic information can be obtained. In fact, much of our more recent knowledge about active and passive drug transport mechanisms has been obtained from studies in various epithelial cell cultures. Both cell culture and transport experiments across epithelial cell monolayers have been automated and these models are now used as a screening tool in drug companies for the production of intestinal drug permeability (Tavelin et al., 2003b).

ABSORPTION IN THE GASTROINTESTINAL TRACT: ROUTES OF PERMEATION ACROSS INTESTINAL CELLS.

Drug transport across the intestinal epithelium may occur by one or more of different routes: passive transcellular and paracellular routes, carrier mediated route, receptor mediated route, active efflux transport and by transcytosis. Several processes limit the permeability: efflux and metabolism in the intestinal cells as well as high molecular weight or unfavorable physicochemical characteristics. Figure 1 shows a scheme of the most important and currently known permeation routes is depicted.

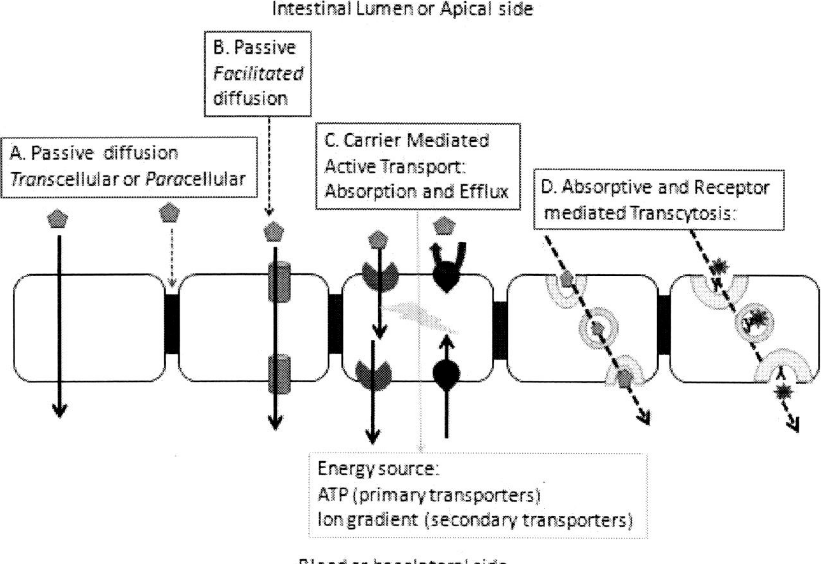

Figure 1. Permeation mechanisms through the intestinal cells.

The transport mechanisms are classified in four groups:

a) Passive diffusion: Driven by the concentration gradient at both sides of the membrane. Transcellular route means that the compound crosses the cell membrane and cytoplasma and paracellular route includes the tight junctions and pores in the membrane.
b) Passive facilitated diffusion (carrier mediated diffusion): it is considered a passive mechanism as it does not require an energy source and the transport is driven by the concentration gradient.
c) Active carrier mediated transport: That can occur in the absorptive (lumen to blood) or the secretive (blood to lumen) direction. In general carrier mediated absorption (active or passive) , is the responsible to transport from intestinal lumen into the blood, hydrophilic and low molecular weight compounds (less than 600Da) (Loscher and Potschka, 2005). Many nutrients such as sugars, aminoacids, small peptides and purine bases use some carrier system to cross the intestinal membrane. AET (active efflux transport) is responsible of the active secretion of a multiplicity of drugs. A consequence, AET substrates cannot effectively penetrate into the body. The best known AET system is P-glycoprotein (Pgp) that limits the transport of a wide range of cationic and lipophilic compounds as cytotoxic anticancer drugs, antibiotics, hormones and HIV protease inhibitors. Other active efflux transporters identified at the intestinal membranes are the MRP proteins (Katsura and Inui, 2003,Loscher and Potschka, 2005).
d) Endocytosis: also called vesicle mediated transport. The molecules or particles are engulfed by a cell. The cell membrane invaginates, encloses the fl uid or particles, then fuses again, forming a vesicle that later transports them across the membrane to the inside of the cell. This mechanism also requires energy expenditure. This

diffusion is dependent on the size of the pore and the molecular weight of the chemical. Phagocytosis is the type of endocytosis where an entire cell is engulfed. Pinocytosis is when the external fluid is engulfed. Receptor - mediated endocytosis occurs when the material to be transported binds to certain specific molecules in the membrane. Classical examples of receptors involved in receptor-mediated transcytosis are the insulin receptor, the transferrin receptor and the transporters for low density lipoprotein, leptin and insulin-like growth factors. Endocytosis probably plays a minor role in drug absorption, except for protein drugs.

In Vitro Models of the Intestinal Barrier: Cell Models

In order to be used as a model on the intestinal barrier the similarities and differences of the cell cultures with the intestinal epithelium related with the main transport routes need to be assessed. For instance, it is necessary to compare, the tightness of the paracellular route, the expression level of transporters or the permeability of the cell membranes.

An overview of cell culture-based models commonly used for permeability assessment can be found in Table 1.

Table 1. Main epitelial cell lines used in drug transport studies

Cell line	Origin	Characteristics	References
CACO-2	Human colonic adenocarcinoma	Well established cell model. Villus enterocyte-like differentiates spontaneously. Expresses some relevant efflux transporters (P-gp, MRP1-2, BCRP…)	(Artursson, 1990,Artursson and Borchardt, 1997,Bleasby et al., 2000,Lesuffleur et al., 1990,Manach and Donovan, 2004)
HT29	Human colonic	Goblet cell-like, forms mucus layer. Crypt cell-like, leaky	(Collett et al., 1996,Ma et al., 1992,Wikman et al., 1993,Wils et al., 1994)
MDCK	Dog kidney epithelium	Low expression of efflux systems, expression systems for transport proteins	(Borst et al., 1999,Cho et al., 1989,Irvine et al., 1999)
LLC-PK1	Pig kidney epithelium	Low expression of efflux systems, expression systems for transport proteins. SuiTable cell line for transfections	(Evers et al., 1996,Miller and Heath, 1989,Tanigawara et al., 1992,Terada et al., 1997)
2/4/A1	Rat fetal intestinal epithelium	Crypt cell-like, leaky, temperature-sensitive differentiation. More suited to study paracellular transport (leakier pores) than CACO-2.	(Paul et al., 1993,Tavelin et al., 1999)
TC7	CACO-2 subclone		(Carriere et al., 1994,Raeissi et al., 1999)

CACO-2

Among the cell culture models, the CACO-2 cell line is the most widely used and best characterized system. Main characteristics are shown in Table 2.

Table 2. Characteristics of parental CACO-2 cells

Origin	Human colorectal adenocarcinoma	(Fliszar et al., 2007,Sugano et al., 2010)
Growth in culture	Monolayer epithelial cells	(Hubatsch et al., 2007,Turco et al., 2010)
Differentiation	14.21 days after confluence in standard culture medium	(Hubatsch et al., 2007,Nomoto et al., 2009,Shimizu et al., 2008,Skolnik et al., 2010,Sugano et al., 2010,Turco et al., 2010)
Morphology	Polarised cells, with tight junctions, apical brush border	(Hubatsch et al., 2007,Skolnik et al., 2010,Turco et al., 2010)
Electrical parameters	High electrical resistance	(des Rieux et al., 2005,Fliszar et al., 2007,Hubatsch et al., 2007,Nomoto et al., 2009,Skolnik et al., 2010,Turco et al., 2010)
Digestive enzymes	Typical membranous peptidases and disaccharidases of the small intestine	(Linnankoski et al., 2008)
Active transport	Amino acids, sugars, vitamins, hormones . .	(Hubatsch et al., 2007,Linnankoski et al., 2008,Skolnik et al., 2010,Sugano et al., 2010)
Membrane ionic transport	$Na+/K+$ ATPase, $H+/K+$ ATPase, $Na+/H+$ exchange, $Na+/K+/Cl^-$ co-transport, apical Cl. channels	(Linnankoski et al., 2008,Nomoto et al., 2009,Skolnik et al., 2010,Sugano et al., 2010)
Membrane non-ionic transporters	Permeability-glycoprotein, multidrug resistant associated protein, lung cancer associated resistance protein	(Corti et al., 2006,Cousein et al., 2007,Fliszar et al., 2007,Linnankoski et al., 2008,Nomoto et al., 2009,Skolnik et al., 2010,Sugano et al., 2010,Zhang et al., 2006)
Receptors	Vitamin B12, vitamin D3, epidermal growth factor, sugar transporters (GLUT1, GLUT3, GLUT5, GLUT2, SGLT1)	(Hubatsch et al., 2007)

Its popularity is based on its human origin and its ability to grow forming polarized monolayers of well differentiated columnar absorptive cells expressing a brush border on their apical surface, some CYP isoenzymes (but not CYP3A), and phase II enzymes, including glutathione S-transferase and sulfotransferase. Despite its cancerous origin, its behavior is similar to its normal in vivo counterpart. Additionally, many active transport systems found in small intestinal enterocytes have been characterized in CACO-2 cells including transport systems for glucose, peptides, aminoacids, bile acids and monocarboxylic acids as well as transport proteins for organic anions and cations, P-gp and MRPs (Bleasby et al., 2000,Delie and Rubas, 1997,Manach and Donovan, 2004). The cells form dense intercellular junctional complexes, resulting in a tight epithelium and thus presenting a unique paracellular as well as transcellular barrier. The use of the CACO-2 cell model permits the investigation of simultaneous absorption routes at the same time (e.g. passive diffusion, active efflux, metabolism), and much insight into the different steps of absorption has been gained by the use of the CACO-2 cell model (Artursson and Borchardt, 1997). Nevertheless, CACO-2 monolayers also present limitations as intestinal membrane model:

1. Paracellular route:

    - More restricted (and similar the the colonic epithelium) due to the higher tightness of the monolayer in comparison with the small intestine epithelium. Average TEER values of this model range between 500-750 $\Omega \cdot cm^2$ (Hidalgo et al., 1989). Cell monolayers with TEER values less than 420 $\Omega \cdot cm^2$ were not used. Lower permeabilities for hydrophilic compounds using the paracellular route that can lead to underestimation of their oral fraction absorbed in humans (Tavelin et al., 2003b).

2. Monolayer composition:

    - Only absorptive cells whereas the human intestinal epithelium included other different cell types as mucus secreting goblet cells.
    - Some authors have proposed the use of co-cultures of HT29-H/CACO-2 cells (Pontier et al., 2001,Walter et al., 1996).

3. Transporter expression levels

    - The potential over expression of P-glycoprotein in pure CACO-2 cells system which may lead to higher secretion rates and consequently lower permeabilities in the absorptive direction (Dahan and Amidon, 2009,Dahan et al., 2009,Shirasaka et al., 2008b).
    - The expression of many uptake transporters is also different in CACO-2 cells, in general they are under expressed compared to the in vivo situation (Anderle et al., 2003,Seithel et al., 2006,Sun et al., 2002,Taipalensuu et al., 2001)

4. Time to confluent monolayers and culture maintenance:

    - Traditional CACO-2 cell culture for permeability studies requires a 3-week period to attain differentiated monolayers.
    - Lentz et al (Lentz et al., 2000) worked with the objective to develop a more rapid, reduced serum culture system for CACO-2 monolayers relative to the traditional 21-day, 10% fetal bovine serum system. They obtained a new model

with 2% iron supplemented calf serum and a combination of growth factors and hormones. The new system provided a rapid, accurate and economical option for passive permeability determination and appears to be applicable to the Biopahrmaceutics Classification System (BCS)

5. Analytical limitations.

   - A major challenge is the development of high-throughput sensitive and accurate methods with a short analysis time to handle the elevated number of samples. This is complicated due to the media composition with high salt content of phosphate buffers and other additives as solubilizers (Balimane et al., 2000).
   - The ideal analytical method should have the speed of a UV-plate reader and the sensitivity of LC/MS and the ability of quantifying simultaneously the drug and marker compounds for paracellular, transcellular and P-gp-mediated efflux, respectively, improving data quality in a more efficient manner (Augustijns and Mols, 2004).
   - Low solubility compounds: The poor aqueous solubility of many new drug candidates further complicate their evaluation in the cell model due to the analytical limitations. Addition of organic cosolvents (DMSO, propylene glycol, N-methyl-2-pyrrolidone, NMP) to the donor solution is a possible way to overcome this problem. However, as the tight junctions are easily compromised, even at low concentrations of cosolvents (<1%), this creates only a limited solution. Inclusion of other solubility improving agents, such as cyclodextrins or bile acids, can be another strategy to work around this issue (Ingels et al., 2004).

6. Hydrodynamic characteristics of the cell system

   - The cell monolayer is a static system in contrast with the dynamic nature of the intestinal environment with pH changes, motility due to muscular contractions, presence of food and secretions.
   - The unstirred water layer is said to be thicker than the one in vivo unless the system is used under agitation with an orbital shaker.

Other Cell Lines

Other cell lines have been used as alternative to CACO-2 cells in order to develop better models to simulate physiological conditions and to optimize experimental conditions by reducing cost and time. The most important are briefly described.

HT29 Cells

Wild-type HT29 cells form multilayers of undifferentiated cells, which are not useful for intestinal permeability studies. Culturing wild-type HT29 cells in media containing galactose instead of glucose, however, leads to the selection of a subclone of HT29 that forms monolayers of polarized cells (Hidalgo et al., 1989). In order to overcome the problems associated to CACO-2 cell monolayers, co-cultures of HT29 and CACO-2 cells have been developed to predict drug permeation. Previous studies have shown that both cell types form monolayers with tight junctions when grown together in culture (Walter et al., 1996,Wikman et al., 1993). The tightness of tight junctions could be modulated by varying cell ratios of CACO-2 and HT29 cells ir order to modify the paracelular permeability of hydrophilic compounds and improve the correlation with human oral fraction absorbed for these compounds (Bohets et al., 2001,Pontier et al., 2001). HT29 cells present a lower expression level of most transporters compared with CACO-2 (Walter et al., 1996,Wikman et al., 1993).

MDCK: Madin Darbin canine kidney cell have received attention as an alternative of CACO-2 cells for permeability measurements. When grown under standard culture conditions, MDCK cells develop tight junctions and form monolayers of polarized cells. The main advantage over CACO-2 cells is the shorter time of confluence (3-5 days). The trans-epithelial electrical resistance of MDCK cells is lower than that of CACO-2 cells and thus, closer to the TEER of the small intestine in vivo. The permeability coefficients of hydrophilic compounds are usually lower in CACO-2 cells than in MDCK cells, which is consistent with the lower TEER values for MDCK cell monolayers (Cho et al., 1989). The nonhuman (canine) and nonintestinal (renal) origin of MDCK cells is considered as a disadvantage. Moreover, they have low expression levels of transporter proteins and low metabolic activity. On the other hand MDCK cells can be easily transfected with P-gp/MDR1 and other transporters and are able to stably express them, leading to the development of transfected, and double transfected cell lines of great utility for mechanistic studies (Adachi et al., 2001,Watanabe et al., 2005).

LLC-PK1: cells derived from pig kidney epithelial cells have been used as an alternative to CACO-2 cells for assessing the permeability (Zhang and Lim, 2008,Zhang et al., 2006).

2/4/A1: The cell model 2/4/A1 originates form fetal rat intestine. Despite it is not from human origin, 2/4/A1 could be considered as a good passive permeation predictor because it lacks functional expression of several important active drugs transporters, including the expression of P-Glycoprotein as efflux transporter (Matsson et al., 2005,Taipalensuu et al., 2001). Although HT29 and CACO-2 co-culture formed leaky tight junctions, 2/4/A1 monolayers present a leakier system, which can better mimic the epithelial barrier to passive drug transport in vivo (Tavelin et al., 2003a). Its average pore radious, 9.0 Åt is quite similar to that reported in the human small intestine (8-13 Å) in contrast with the smaller pores in CACO-2 cells (<6 Å) (Fine et al., 1995,Tavelin et al., 2003a). On the other hand monolayer's leakiness can be modified, changing the temperature in which cell culture is growing from 33°C to 39°C and the modification of the paracellular pathway could allow its use in other physiological barriers as blood-brain barrier, where the paracellular via is tighter. As paracellular pathway is a significant contributor to total drug permeability for many incompletely absorbed, low permeability compounds the need to include the paracellular pathway in models used for drug permeability screening has been proposed especially when are expected low-permeability compounds (Artursson, 1990,Artursson et al., 2001,Matsson et al., 2005).

In conclusion, due to the advantages described above, and shorter cultivation time (4-6 days compared to 21 days for CACO-2 cells using the standard cell culture procedure), 2/4/A1 cell line could be an optional tool in permeation studies where the objectives are slow and incomplete absorbed drugs that are significantly transported via the paracellular route such as atenolol (Adson et al., 1994,Collett et al., 1996), H2 antagonists (Gan et al., 1998,Lee and Thakker, 1999), furosemide (Flanagan et al., 2002,Pade and Stavchansky, 1997) and hydrophilic peptides (He et al., 1998,Knipp et al., 1997,Pauletti et al., 1997).

TC-7: Clonal selection has been used to obtain CACO-2 cells with stronger metabolic competence. The TC7 subclone has been characterized for its ability to transport and metabolize endogenous compounds and xenobiotics. Hence, the TC-7 clone expresses similar levels of sucrose isomaltase compared to human jejunum, while levels are much lower in parental CACO-2 (Pfrunder et al., 2003); enzymes involved in phase II metabolic reactions are also present in the TC-7 clone as in human jejunum, while their activity is very low in CACO-2 cells (Gres et al., 1998). Parental CACO-2 cells express only CYP3A5 (Engman et al., 2001,Raeissi et al., 1999). It was therefore suggested that the metabolic competence of TC7 cells makes this clone a valid alternative for the parental cell line particularly suiTable for drug transport and biotransformation studies (Le Ferrec et al., 2001). However, the degree of expression of CYP3A in TC7 cells still remains lower than the expression level in human enterocytes or mucus-secreting goblet cell.

CACO-2 Cells for Predicting Oral Fraction Absorbed: Comparison with Human, Rat and PAMPA Models

In general the correlations between permeability values obtained with different experimental systems (as CACO-2, rat perfusion techniques or human intestinal perfusion) and the oral fraction absorbed in humans (or rats) are sigmoidal curves with a steep slope defining the change in fraction absorbed with permeability. The sigmoidal curves obtained with the different models do not overlap but are separated due to the different magnitude of the permeability values of each experimental system as it can be seen in Figure 2 (Lennernas, 1998,Sanchez-Castano et al., 2000).

The differences in magnitude are explained base on the differences in surface available for diffusion, the different paracellular permeability and the differences in the unstirred water layer. Several authors have attempted to model these differences in order to be able to scale up a CACO-2 permeability value to a human permeability one. In general it has been accepted that the in vitro model has a more restricted paracellular pathway, and that high and low permeability drugs have different accessibility to the intestinal surface (with highly permeable compounds using only the villus tips while low permeability drugs use the whole mucosal surface). Nevertheless a more recent biophysical model approach showed different conclusions supporting the hypothesis, based on their modeling approach, that the unstirred water layer thickness in the human in situ experiments is much greater than had been commonly thought, the absorption-accessible mucosal surface area in humans is apparently not dependent on lipophilicity of the drug, in contrast to rodent studies, and the relative "leakiness" of the human jejunumis not so different from that observed in a number of CACO-2 studies (Avdeef, 2010).

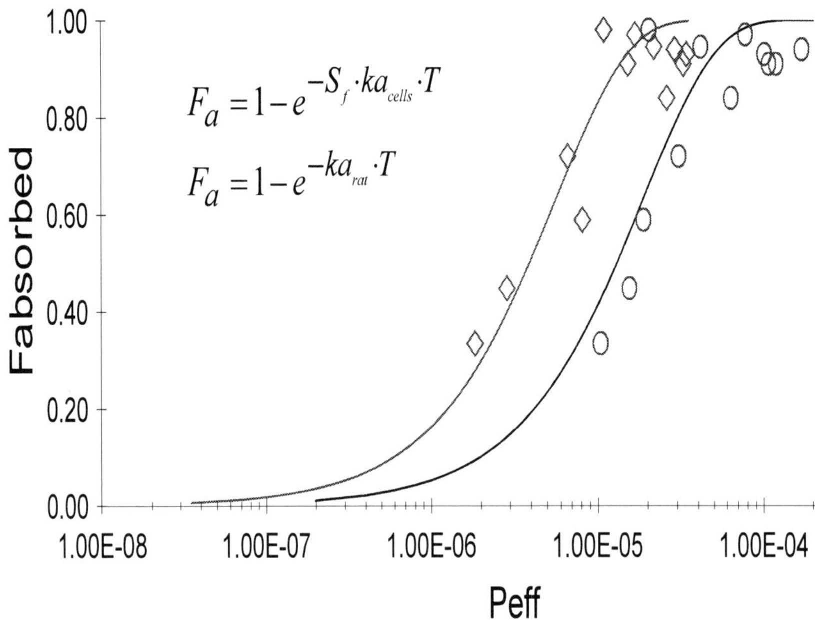

Parameter	Units	Estimate	StdError	CV%
T	hours	**1.314**	0.175	13.32
SF	-	**5.587**	1.118	20.02

Figure 2. Correlation of permeabilities obtained in rat (circles) and CACO-2 (diamonds) and oral fraction absorbed in rat. Parameters of the model to fit simultaneously both data sets where permeabilities are represented by the absorption rate coefficients (ka=(Area/Volume)*Peff).

In Table 3 a summary of the correlations attempted between different experimental models are described.

Our research group has studied in detail the absorbability of a family of fluoroquinolonas in different experimental systems in order to compare their predictive performance of oral fraction absorbed. In Figure 3 the correlation between permeability in rat intestine (obtained with Doluisio's perfusion method) or permeability in CACO-2 cells with oral fraction absorbed in rats is shown.

As it can be seen a similar correlation was obtained with both absorption model and the main difference is the magnitude of the permeability values that is higher in the rat model. As a first attempt to explain this difference it was considered that the paracellular permeability of this family of quinolones was small and comparable in both systems and the main difference could be attributed to the differences in surface area available for transport as in the CACO-2 cells the folds and villi of the intestine are not present.

Table 3. Correlations attempted between different experimental models are described

Variables correlated	Model drugs used	Main conclusions	References
CACO-2 peff versus fa human	Passively absorbed compounds	Sigmoidal correlation between Peff and Fa but very steep so small variations in permeability are reflected in high variations in Fa	(Amidon et al., 1995,Artursson and Karlsson, 1991,Conradi et al., 1993,Kim et al., 1993,Lennernas et al., 1997,Walter et al., 1995)
Caco-versus human jejunum peff	Passively absorbed compounds and carrier mediated	Good correlation for passively absorbed compounds Differences in magnitude (Human> CACO-2; 2 to 4 fold for high permeability drugs and 20 to 80 folds for low permeable drugs) Correlation not unbiased by unstirred water layer Hypothesis that highly permeable compounds use only the surface of villus tips for being absorbed. Poor correlation for actively absorbed drugs Problems due to high variability of the technique intra and inter-laboratory	(Artursson and Karlsson, 1991,Conradi et al., 1993,Hu et al., 1988,Jumarie and Malo, 1991,Kim et al., 1993,Lennernas et al., 1993,Lennernas and Regardh, 1993,Stewart et al., 1995,Sugano et al., 2010,Vachon and Beaulieu, 1992,Volpe, 2010,Walter and Kissel, 1994,Walter et al., 1995,Wilson and Dietschy, 1974,Yamaguchi et al., 1986)
Peff human versus Fa human		"Gold standard".	(Amidon et al., 1995)
CACO-2 versus animal perfusion data			(Fernandez-Teruel et al., 2005,Gonzalez-Alvarez et al., 2007,Gonzalez-Alvarez et al., 2009,Rodriguez-Ibanez et al., 2003,Rodriguez-Ibanez et al., 2006)

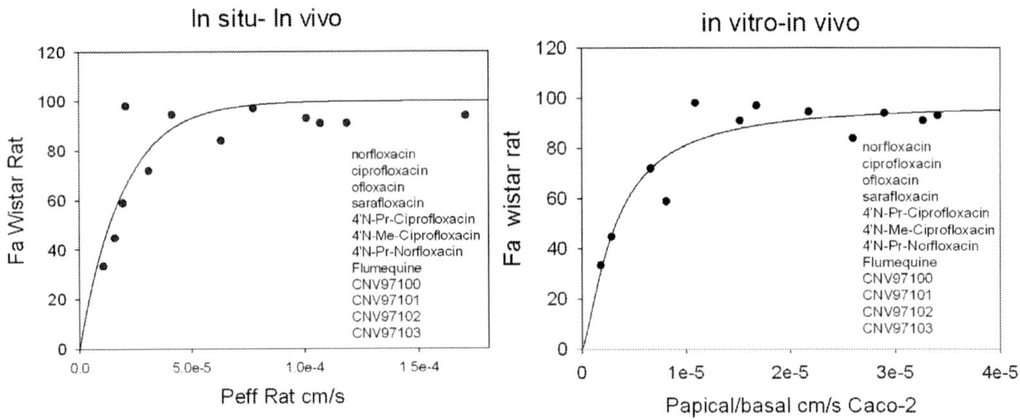

Figure 3. Correlation between permeability and fraction absorbed obtained in rat and cell model of different fluoroquinolones.

In the Figure 2 both data sets are depicted together and the fit to a single model with common parameters for both absorption techniques. There is only one parameter for scaling up from CACO-2 to rat data, Sf, as the coefficient to account for the difference in the surface area available for transport, so in this simple model it is assumed that paracellular permeability is the same in both systems as well the unstirred water layer resistance. This correlation of permeabilities and oral fraction absorbed is based on a previous one developed by our group (Sanchez-Castano et al., 2000). T represents the absorbing time in the rat small intestine. Absorption rate coefficients obtained in situ in rats were transformed into permeabilities with the following equation Peff= ka*R/2 where R is the effective radious of the intestinal segment.

On the other hand in order to transform CACO-2 permeabilities into absorption rate coefficients considering the area of the insert and the volume of the donor chamber with this relationship: ka_{cells}=(Area/Volumen)*Peff.

These results of fluoroquinolone absorption in rat and CACO-2 cells have been also correlated with data obtained using PAMPA membranes (Bermejo et al., 2004). The results are depicted in Figure 4.

So, when the intrinsic PAMPA data (*P*o) are corrected for the appropriate unstirred water layer permeability ($P\text{bio}_u$), the unified relationship between the biological barrier data ($P\text{bio}_{app}$) and the artificial membrane data is

$$\log P_{app}^{bio} = -0.362 + 0.886\left[\log P_0 - \log\left(1 + \frac{P_0}{P_u^{bio}}\right)\right] \quad \text{Equation 1}$$

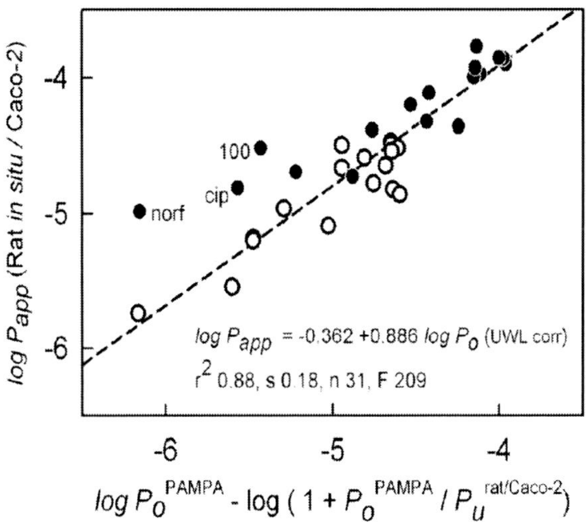

Figure. 4 The relationship between UWL-corrected PAMPA data and the combined rat in situ and CACO-2 results. Taken from (Bermejo et al., 2004).

In conclusion PAMPA measurements can predict quite well fluoroquinolone in situ and in vitro permeability data with the advantage of being a highthrougput system more adequate for rapid screening of families of compounds. On the other hand it has to be considered that the data used for correlation with PAMPA results in this example were permeabilities of compounds that are transported by passive diffusion or in the case of those quinolones with a carrier (absorptive or secretive) involved in their transport the diffusional permeability value was the one used for the correlation (this value was estimated either by determining the permeability under saturating conditions or with a selective inhibitor of the carrier). In summary the PAMPA model is very useful for predicting the passive diffusion component but rat or CACO-2 model are necessary for characterizing carrier mediated processes.

In summary, the results obtained to date indicate that CACO-2 monolayers can be used to characterize drug transport by different pathways across the intestinal epithelium but the best correlations with the absorbed fraction in vivo are obtained when only passively transported drugs are included as the differences in the expression level of the transported from in vitro to the in vivo situations preclude a direct comparison of in vitro and in vivo data unless a good scaling factor is incorporated in the model (Artursson et al., 2001).

CACO-2 Cells for the Study of Drug Absorption Mechanisms: Modeling Approaches for Parameter Estimation

Apparent Permeability Estimation (Sink versus Non Sink Conditions)

In the CACO-2 system if the studied compound is transported by passive diffusion, the transport is driven by the concentration gradient in both sides of the monolayer. It is non saturable so the permeability value should be independent of the concentration and of transport direction (if the pH in both chambers is the same). In the case of a carrier mediated process it is necessary to model the changes of permeability with concentration in order to delineate the diffusion component of the transport and the parameters of the carrier mediated process i. e Km and Vmax. The first step, in consequence, in any of the above scenarios is to estimate the apparent permeability value from the raw experimental data i.e from the concentrations of drug in the receiver chamber. This point is briefly analyzed below as the assumptions underlying the usual calculation methods are often overlooked.

The diffusion process through the CACO-2 monolayer can be described using Fick's first law:

$$\frac{dM}{dt} = A \cdot Peff \cdot (C_{donor} - C_{acceptor}) \qquad \text{Equation 2}$$

where Peff is the effective or apparent permeability, A is the transport area (the surface of the insert) and C_{donor} and $C_{acceptor}$ are the concentrations in the donor and acceptor chamber and dM/dt represent the amount or mass of drug accumulated with time in the acceptor or receiver side. Under sink conditions $C_{acceptor}$ is negligible versus C_{donor} and in this case permeability is estimated from the slope of the linear regression of accumulated amounts of drug in the acceptor chamber divided by the area and the initial concentration in the donor one.

$$\frac{dM}{dt} = A \cdot Peff \cdot C_{donor} \qquad \text{Equation 3}$$

$$\frac{dM}{dt} \approx \frac{\Delta M}{\Delta t} = A \cdot Peff \cdot C_{donor} \qquad \text{Equation 4}$$

$$Peff = \frac{\frac{\Delta M}{\Delta t}}{A \cdot C_{donor}} \qquad \text{Equation 5}$$

In general it is accepted that sink conditions are maintained when $C_{acceptor}$ is less than 10% of C_{donor}.

The above approach is an *approximation* as we are considering increments instead of differentials. On the other hand, it lies under the assumption that there is no compound retention in the cell layer or the plastic ware and also implies that the amount (and concentration) of drug in the donor compartment does not change during the time of the experiment.

Nevertheless, the amount of drug in donor concentration decreases over time and even when sink conditions are kept ($C_{acceptor} < 0.1 \cdot C_{donor}$) a more accurate estimation of Peff can be obtained if the amount in donor (and then donor concentration) is corrected for the loss at each sampling point. In that case permeability is estimated from the slope of the linear regression on the "accumulated fraction transported" versus time. These accumulated fractions are calculated from the amount M transported in each time interval (M in receiver at time (t-1)- M receiver at time (t)) over the surface area (A) and the average concentration in the donor at the beginning and at the end of each time interval, from equation 5

$$\text{fraction transported in each time interval} = \frac{\Delta M}{A \cdot \overline{C}_{donor}} = Peff \cdot \Delta t \qquad \text{Equation 6}$$

In Figure 5 the estimated permeability value with both approximations is depicted under sink and non sink conditions. As it can be observed if sink conditions are not maintained the difference between both estimations is bigger. Then, the question is which of both values is more accurate?

The integral equation describing the evolution of the concentrations in the receiver with time is equation 7 that is the true solution to the differential equation 3. This equation is applicable with sink and non sink conditions as in the case of sink conditions the result is closed to the obtained with the approximation described with equation 5 and 6, but in the case of non sink conditions equation 7 will provide the most accurate result.

$$C_{R(t)} = \left(\frac{M_{tot}}{V_D + V_R}\right) + \left(C_{R0} - \frac{M_{tot}}{V_D + V_R}\right) \cdot e^{-P_{eff} \cdot A \cdot \left(\frac{1}{V_D} + \frac{1}{V_R}\right) \cdot t} \qquad \text{Equation 7}$$

where M_{tot} is the total amount of drug in the system, V_D and V_R are the donor and receiver volume respectively, C_{R0} is the concentration in receiver at time 0 (in general $C_{R0}=0$), A the surface of the monolayer, t the time and Peff the effective permeability.

This equation is of application to an undisturbed system, but when a sample is taken the receiver concentration is diluted, so some corrections are necessary to estimate Peff. The permeability value with this equation is estimated by non linear regression of receiver concentrations versus time. A detailed description of the derivation of this equation and of the calculations has been published by Tavelin S et al (Tavelin et al., 2002). The relative errors in permeability estimation in the two previous examples in comparison with the non linear regression equation are shown in Table 3.

f=dilution factor= (Vr-Vs/Vr)

Area cm2	1.13	Donor Volume (Vd) mL	0.40	f	0.50
Receiver Volume (Vr) mL	1.20	Sample Volume (Vs) mL	0.60		
		Initial Conc. Donor μM	500.00	Initial amount donor nM	200.00

time s	Conc experimental receiver μM	Amount in receiver nMol	Amount in sample nMol	Accumulated Amount in sample nMol	Accumulated Amount in receiver nMol	Conc. Theor Donor μM	Crel	sink?	Accumulated fraction transported
0	0	0.0	0.0	0.0	0.0	500.0			0.000
600	21.80	26.2	13.1	13.1	26.2	434.6	0.050	ok	0.050
1200	29.35	35.2	17.6	30.7	48.3	379.2	0.077	ok	0.098
1800	30.52	36.6	18.3	49.0	67.3	331.7	0.092	ok	0.145
2400	29.00	34.8	17.4	66.4	83.8	290.5	0.100	ok	0.192
3000	26.45	31.7	15.9	82.3	98.1	254.6	0.104	ok	0.238
3600	23.66	28.4	14.2	96.5	110.7	223.3	0.106	ok	0.285
		Creceiver*Vr	Creceiver*Vs		Mreceiver+ Msamples	(Mo-Mreceiver) / Vd	Creceiver/ Cdonor		accum ΔM/(C'·A)

	SINK analysis	SINK analysis Corrected by Cdonor
Slope	0.0280	
intercept	13.54	0.002
r	0.995	1.000
Peff cm/s	4.96E-05	7.89E-05

C' average concentration between two sampling points

accum ΔM =calculated amount transported in each time interval and accumulated

Figure 5A. Estimation of permeability value from CACO-2 experimental data. Sink conditions are kept in this example.

f=dilution factor= (Vr-Vs/Vr)

Area cm2	1.13	Donor Volume (Vd) mL	0.40	f	0.50
Receiver Volume (Vr) mL	1.20	Sample Volume (Vs) mL	0.60		
		Initial Conc. Donor µM	50	Initial amount donor nM	20

time s	Conc experimental receiver µM	Amount in receiver nMol	Amount in sample nMol	Accumulated Amount in sample nMol	Accumulated Amount in receiver nMol	Conc. Theor Donor µM	Crel	sink?	Accumulated fraction transported
0	0	0.00	0.00	0.00	0.00	50.00			0.000
600	2.97	3.57	1.78	1.78	3.57	41.09	0.07	ok	0.069
1200	3.83	4.60	2.30	4.08	6.38	34.05	0.11	No sink	0.135
1800	3.83	4.59	2.30	6.38	8.67	28.32	0.14	No sink	0.200
2400	3.48	4.18	2.09	8.46	10.55	23.62	0.15	No sink	0.264
3000	3.03	3.64	1.82	10.28	12.10	19.75	0.15	No sink	0.328
3600	2.60	3.12	1.56	11.84	13.40	16.49	0.16	No sink	0.391
		Creceiver*Vr	Creceiver*Vs		Mreceiver+ Msamples	(Mo-Mreceiver) / Vd	Creceiver/ Cdonor		accum ΔM/(C''·A)

	SINK analysis	SINK analysis Corrected by Cdonor
Slope	0.0032	
intercept	2.29	0.003
r	0.990	1.000
Peff cm/s	5.75E-06	1.08E-04

C'' average concentration between two sampling points

accum ΔM =calculated amount transported in each time interval and accumulated

Figure 5. B. Estimation of permeability value from CACO-2 experimental data. Sink conditions are not maintained in this example.

Table 3. Relative errors in permeability estimation in comparison with the non linear regression equation

Example 1	SINK	SINK	No SINK
Slope	0.0280	Corrected by Cdonor	Solver estimation
intercept	13.54	0.002	
r	0.995	1.000	1.000
Peff cm/s	4.96E-05	7.89E-05	**8.45E-05**
%Relative error	-41.34	-6.61	
Slope	0.0032	Corrected by Cdonor	Solver
intercept	2.29	0.003	
r	0.990	1.000	1.000
Peff cm/s	5.75E-05	1.08E-04	**1.20E-04**
%Relative error	-52.03	-9.60	

When non sink conditions are clearly violated by far more than 20% of C donor at last sampling time, then estimation error could be higher than 25%.

Mathematical Modeling of CACO-2 Data for Parameter Estimation

In order to detect if a carrier mediated process is involved in the transport of the substance is advisable to perform permeability experiments at different initial concentrations in both directions. The use of pH gradient in both chambers is not adequate in particular for ionizable compounds to avoid differences in apical to basal and basal to apical permeabilities due to the ion trapping effect. If there are statistical differences in the apparent permeabilities with the concentration changes and the use of inhibitors also affect the apparent permeability, then it is reasonable to suspect the contribution of a transport protein and it is possible to characterize the components of the transport by different modeling approaches:

1) Modeling the change of apparent permeabilities versus concentration: *P versus C*
2) Kinetic compartmental modeling of the changes of amounts versus times in donor, receiver and cell compartment: *M versus t*

These methods are illustrated with two examples:

P versus C

In the following Figure the apical to basal and basal to apical permeabilities of Grepafloxacin in CACO-2 cells obtained at four initial donor concentrations are represented.

Apical permeability increases and basal values decrease as concentration is raised from 25 to 2500 μM. The ratio basal/apical permeability is higher than one for the three lower concentrations, and the secretory system is probably saturated at the highest concentration—i.e. the ratio basal/apical permeability equals one. We have previously demonstrated in CACO-2 cells that the transporter(s) are sensitive to verapamil, quinidine and ciclosporin A (Rodriguez-Ibanez et al., 2003).

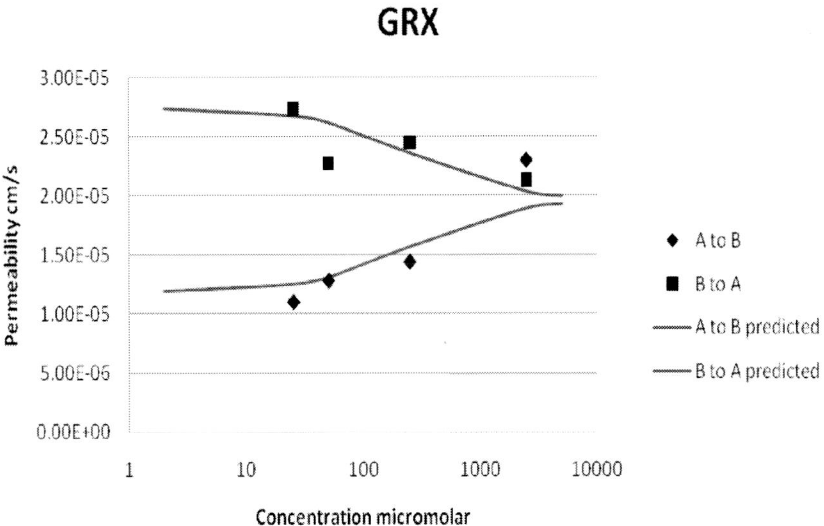

Figure 6. Concentrations in acceptor chamber different initial concentrations in both directions. B-A: basolateral to apical experiments and A-B apical to basolateral experiments.

In order to estimate the transport parameters apical to basal permeabilities and basal to apical ones where used to fit these equations simultaneously:

$$P_{eff_{a-b}} = P_{diff} - \frac{Vm}{(Km + C)} \quad \text{Equation 8}$$

$$P_{eff_{b-a}} = P_{diff} + \frac{Vm}{(Km + C)} \quad \text{Equation 9}$$

where Pdiff represents the diffusion component and Vm and Km are the Michaelis Menten parameters

The estimated parameters where:

Pdiff= 1.96 e-5 cm/s; Km=259.46 μM and Vm=2.03e-3 μMol/(cm^2·s).

In this model there are several assumptions, for instance, there are only two compartments apical and basolateral so the membrane is a volume less barrier and the main limiting step for the transport is the membrane (and the carrier) so the unstirred water layer effect is negligible.

In overexpression cell systems, as transfected cell lines with some particular transporter gen i.e MDCK-Mdr1 unstirred water layer resistance could bias the estimation of Km, where the extent of bias is determined by transporter expression level and substrate affinity (Balakrishnan et al., 2007).

If the unstirred water layer resistance is considered the total resistance Rtotal to the flux of the drug is

$$R_{TOTAL} = R_{UWL} + R_{MONOLAYER}$$ Equation 10

Since permeability is the inverse of the resistance

$$\frac{1}{Peff} = \frac{1}{Puwl} + \frac{1}{Pcells}$$ Equation 11

where Peff is the effective or apparent permeability, Pcell the permeability through the cell monolayer and Puwl the permeability in the unstirred water layer.

The permeability through the cell monolayer in the a to b direction in the case of the presence of a secretion carrier could be described with equation 8:

$$P_{cell} = P_{diff} - \frac{Vm}{(Km + C)}$$ Equation 12

where Pdiff is the passive diffusion component. Substituting equation 12 into equation 11:

$$Peff_{\text{a-b}} = \frac{Puwl \cdot (Pdiff - \frac{Vm}{Km + C})}{Puwl + (Pdiff - \frac{Vm}{Km + C})}$$ Equation 13

and consequently,

$$Peff_{\text{b-a}} = \frac{Puwl \cdot (Pdiff + \frac{Vm}{Km + C})}{Puwl + (Pdiff + \frac{Vm}{Km + C})}$$ Equation 14

In order to fit both equations simultaneously to the a to b and b to a permeabilities it is necessary to have an estimation of the Puwl that could be obtained from the permeability of the compound in the system (inserts) without monolayer.

An example of this model and the potential bias in estimation of carrier parameters is described in Balakrishnan et al (Balakrishnan et al., 2007).

Amounts versus Time

Gonzalez-Alvarez et. al. (Gonzalez-Alvarez et al., 2005), have used a three compartment model approach to elucidate the secretion pump location of a new fluoroquinolone (CNV97100), the binding site of the transporter to the substrate as well as the relevance of the secretion processes observed in two CACO-2 cell line cultures (ATCC and TC-7) with a slight difference in P-gp expression levels.

In Figure 7 the apparent permeability values obtained in the ATCC CACO-2 cell line in both directions and at several initial concentrations as well as in presence of verapamil are shown. Similar results were obtained in the TC-7 clone.

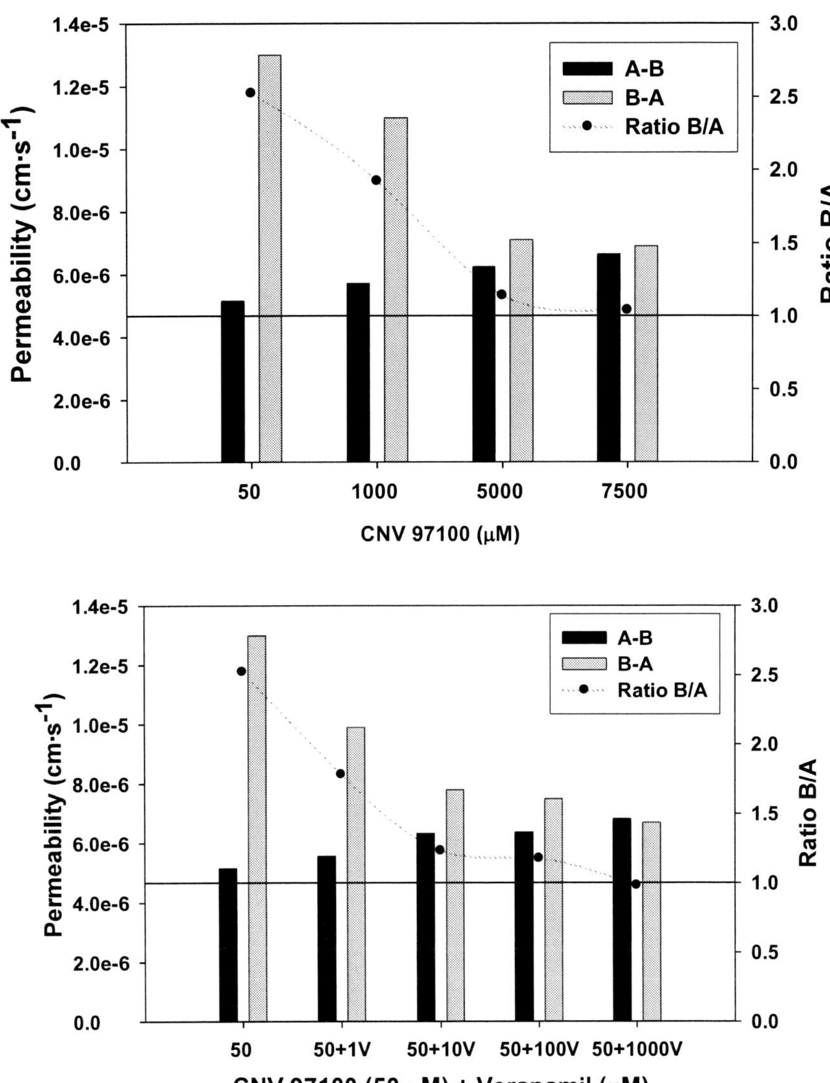

Figure 7. a) Permeability and efflux ratios (P_{BA}/P_{AB}) of CNV 97100 at different concentrations. The experiments were performed at 37°C. Each bar is the average of three determinations. b) Permeability and efflux ratios (P_{BA}/P_{AB}) showing effects of Verapamil on the transport of CNV97100.

In Figure 8 the experimental and predicted accumulated concentrations in receiver chamber at all the initial concentrations and in both directions and cell line are shown.

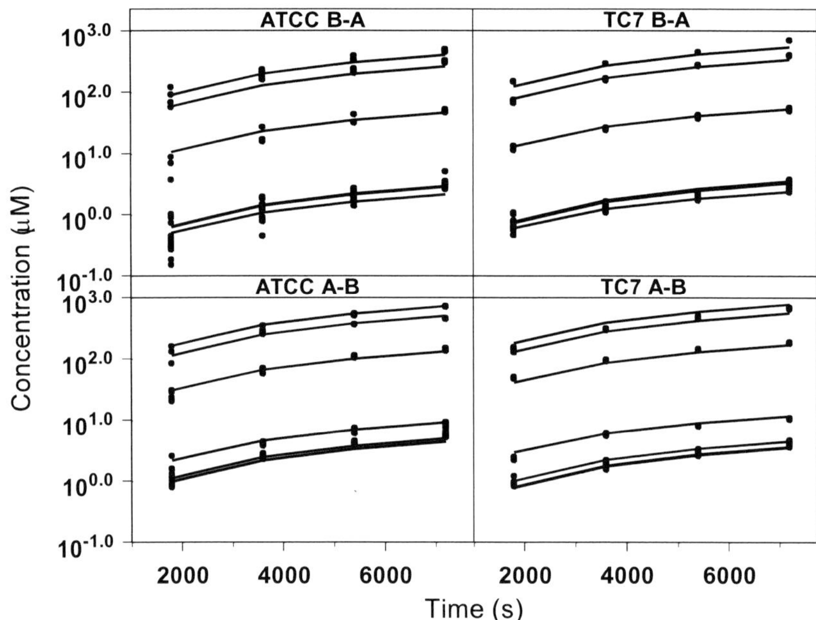

Figure 8. Concentrations in acceptor chamber versus time at different initial concentrations and in presence and absence of Verapamil in both cell lines. B-A: basolateral to apical experiments. A-B apical to basolateral experiments.

The differential equations for the apical, cell and basolateral compartment where:

$$\frac{dQ_A}{dt} = Cl_{AC} \cdot \left(\frac{Q_C}{V_C} - \frac{Q_A}{V_A} \right) + \frac{Vm \cdot \frac{Q_C}{V_C}}{Km + \frac{Q_C}{V_C}}$$ Equation 15

$$\frac{dQ_C}{dt} = -Cl_{CB} \cdot \left(\frac{Q_C}{V_C} - \frac{Q_B}{V_B} \right) - Cl_{AC} \cdot \left(\frac{Q_C}{V_C} - \frac{Q_A}{V_A} \right) - \frac{Vm \cdot \frac{Q_C}{V_C}}{Km + \frac{Q_C}{V_C}}$$ Equation 16

$$\frac{dQ_B}{dt} = Cl_{CB} \cdot \left(\frac{Q_C}{V_C} - \frac{Q_B}{V_B} \right)$$ Equation 17

where dQA/dt, dQC/dt and dQB/dt are the rates of change of drug amounts in the apical, cellular and basal compartment, respectively, Vm and Km are the Michaelis–Menten parameters, ClAC and ClCB are the clearances apical to cell and cell to basolateral, respectively. QA, QB and QC are the drug amounts over time in the apical, cellular and basolateral chamber, respectively; and VA, VB and VC are the volumes of each compartment.

The models used to describe the transepithelial transport processes are shown in Figure 9.

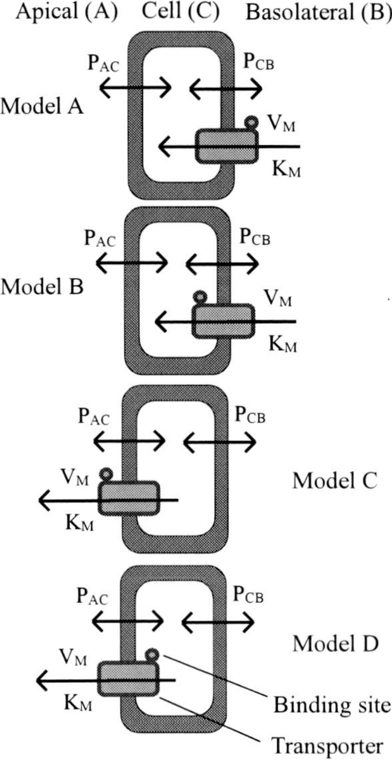

Figure 9. Model A to D with different location on the transporter and binding site for the substrate. Mode D was selected as the best one and it considers an efflux transporter located at the apical membrane with intracellular binding site, a common K_M in cell lines, different V_M parameters in ATCC versus TC7 and different permeability of the apical and basolateral membrane.

In order to calculate the permeability in each side (i.e the permeability of the apical membrane of the enterocite and the permeability of the basolateral membrane) the following relation was used: P=Cl/S where Cl is the clearance, P is the permeability and S is the monolayer cell surface, which was the same for all the inserts and had a value of 4.2 cm2. The experimental and fitted values are represented in Figure 10.

The models were constructed based in the two possible locations of the efflux pump, apical or basolateral, and the two possible binding sites for the quinolone, intra- or extracellular. The rates of change of drug amounts in each compartment (apical, cellular and basolateral) dQ/dt where described in terms of differential equations.

All the models and differential equations were fitted to the whole data set (amounts in donor and receiver chambers in both directions and at the different initial concentrations. The model with the best fit criteria (minimum objective function) was selected.

The final best selected model was model D: apical location of the transporter with intracelular binding site (or intra-membrane) which is consistent with the accepted characteristics for P-gp.

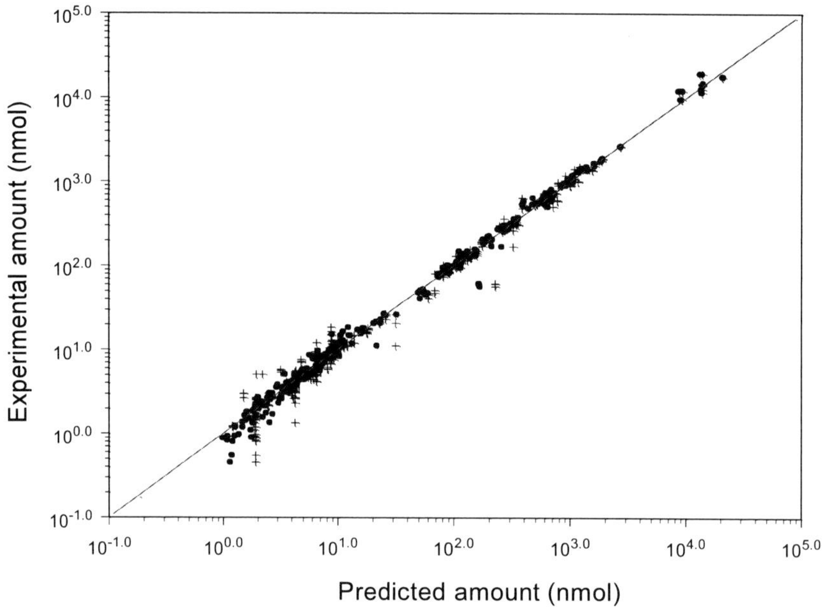

Figure 10. Experimental versus predicted concentrations using Model D which considers an efflux transporter located at the apical membrane and intracellular binding site, a common K_M in cell lines, different V_M parameters in ATCC versus TC7 and different variability for the diffusion component.

The advantage of this procedure is that it is applicable for linear and non-linear transport of drug with time, and it can be used even if the concentration in the receiver chamber exceed by more than 10% the donor one i.e. the method can be applied when sink conditions are not achieved. It accommodates negligible or relevant back flow, and would allow the adequate correction in the case of relevant mass balance problems on the other hand by model selection procedures it and allows the evaluation of the location of the transporter and its binding site. On the other hand it requires a more intensive sampling scheme (with samples in both donor and receiver chambers) and software for non-linear regression.

The parameters obtained with this approach are shown in the next Table 4.

Table 4. Parameters obtained for the best pharmacokinetic model by a simultaneous fit between ATCC and TC7 cells

Parameter	Values (CV%)	
	ATCC	TC7
Cellular volume (mL)	$7.38 \cdot 10^{-2}$ (22.2)	
K_M (μM)	$3.76 \cdot 10^{2}$ (37.5)	
P_{AC} ($cm \cdot s^{-1}$)	$8.40 \cdot 10^{-6}$ (7.5)	
P_{CB} ($cm \cdot s^{-1}$)	$3.45 \cdot 10^{-5}$ (28.8)	
V_M ($nmol \cdot s^{-1} \cdot cm^{-2}$)	$2.59 \cdot 10^{-2}$ (24.6)	$5.32 \cdot 10^{-2}$ (22.4)

While the parameters obtained with the classical approach (Peff versus concentration) are the following ones:

Table 5. Parameters of the two-compartment fit

Parameter	Value	CV(%)
P_{diff} ($cm{\cdot}s^{-1}$)	7.08×10^{-6}	2.96
V_M ($nmol{\cdot}s^{-1}{\cdot}cm^{-2}$)	6.14×10^{-3}	41.70
*ΔV_M ($nmol{\cdot}s^{-1}{\cdot}cm^{-2}$)	2.13×10^{-3}	76.08
K_M (μM)	1483.5	46.31
K_I (μM)	0.80	53.22

*Corresponds to the increase in Vm in the TC7 cell line due to higher transporter expression.

In this case Vm is underestimated while Km is overestimated with respect to the three compartment model because this later one considers the concentration that actually the carrier "sees" (the intracellular one). This aspect is in particular relevant when the parameter data is going to be used in quantitative structure activity relationships.

A similar approach of parameter estimation by curve fitting of the differential equations describing the evolution of amounts versus time was used to explore the absorption mechanism of sarafloxacin (Fernandez-Teruel et al., 2005). In this case the curve fitting and model selection procedure allowed to detect the presence of two transporters (one of absorption and one of secretion) and to establish a rational hypothesis about its location.

Correlation between CACO-2 Cells Data and in Situ Rat Data for the Study of P-GP Mediated Secretion

When the differences in expression level of the carrier are not very big it is possible to model globally rat and CACO-2 data even for compounds with a carrier mediated process (as p-glycoprotein secretion) as it can be seen in the next example.

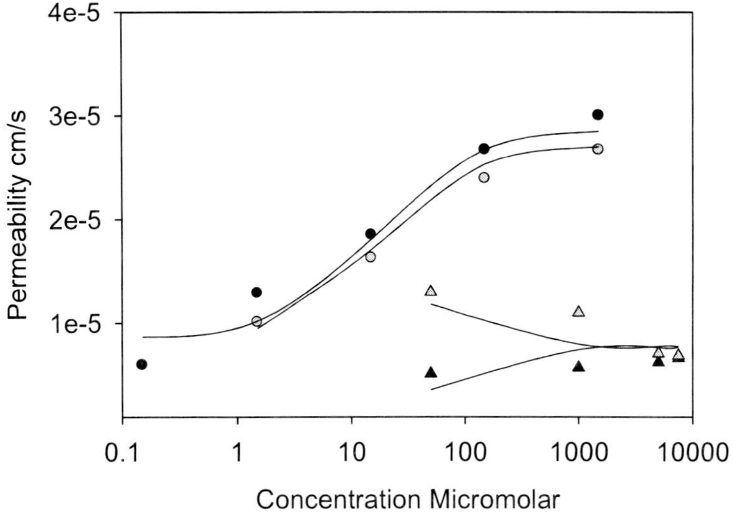

Figure 11. Experimental permeability values in whole small intestine (black circles) and ileum (grey circles) and in CACO-2 cells (A to B, black triangles) and (B to A, grey triangles). The lines correspond to the fit predicted values. Adapted from Gonzalez et al.(Gonzalez-Alvarez et al., 2005).

Figure 11 represents rat in situ permeability data of a fluoroquinolone (CNV97100) obtained a different initial concentrations in the whole small intestine and in ileum of rats. In the same Figure the apical to basal and basal to apical permeabilities in CACO-2 at different initial concentrations are also represented.

A mathematical model was fitted to the whole data set considering the same transport parameters (same Km and Vmax for the carrier) and correcting for the differences in surface (Sf) and the expression level (Ef). In the case of the in situ data, the surface factor was corrected by the difference in the effective radious in whole small intestine versus ileum and the Vmax value in whole small intestine was corrected to incorporate the slightly higher expression level of P-glycoprotein in ileum compared with the average expression in the whole small rat intestine. The equations of the model are the following.

$$\textit{Apical to Basal} \quad P_{eff} - P_{diff} - \frac{E_f \cdot Vm}{Km + C} \qquad \text{Equation 18}$$

$$\textit{Basal to Apical} \quad P_{eff} - P_{diff} + \frac{E_f \cdot Vm}{Km + C} \qquad \text{Equation 19}$$

$$\textit{Rat} \quad P_{eff}^{rat} - S_F \cdot \left(P_{diff} - \frac{Vm}{Km + C} \right) \qquad \text{Equation 20}$$

Table 6. Parameters of CNV97100 in vitro-in situ correlation

Parameter	Estimate	SE	CV(%)
P_{diff} ($cm \cdot s^{-1}$)	7.71×10^{-6}	5.58×10^{-7}	7.76
V_M ($nmol \cdot s^{-1} \cdot cm^{-2}$)	9.06×10^{-5}	2.82×10^{-5}	31.15
K_M (μM)	16.53	5.50	33.29
S_F	3.53	0.30	8.57
E_f	3.02	1.13	37.49
R^2		0.965	
SSR		4.01×10^{-11}	
AIC		394.98	

As it can be seen in the Table 6 with the parameters obtained after fitting the model, the surface factor Sf indicates that the passive permeability of CNV97100 differs between the two systems 3.5-fold, in agreement with previous studies (Stewart et al., 1995). Additionally, the model estimates a 3-fold difference in expression level between CACO-2 cells and rat small intestine (represented by Ef). The mathematical model used to correlate the *in vitro* and *in situ* permeabilities has been constructed using very simplistic assumptions and it would require further refinement but it is an example of how identifying the physiological and anatomical differences in the models and with the right scaling factors CACO-2 cells are a good model for predicting human oral absorption (Avdeef and Tam, 2010).

A very similar approach has been used to correlate grepafloxacion permeabilities obtained in rat and in CACO-2 cells. In the next Figure (Figure 12) the experimental versus predicted permeabilities values in rat and CACO-2 are depicted.

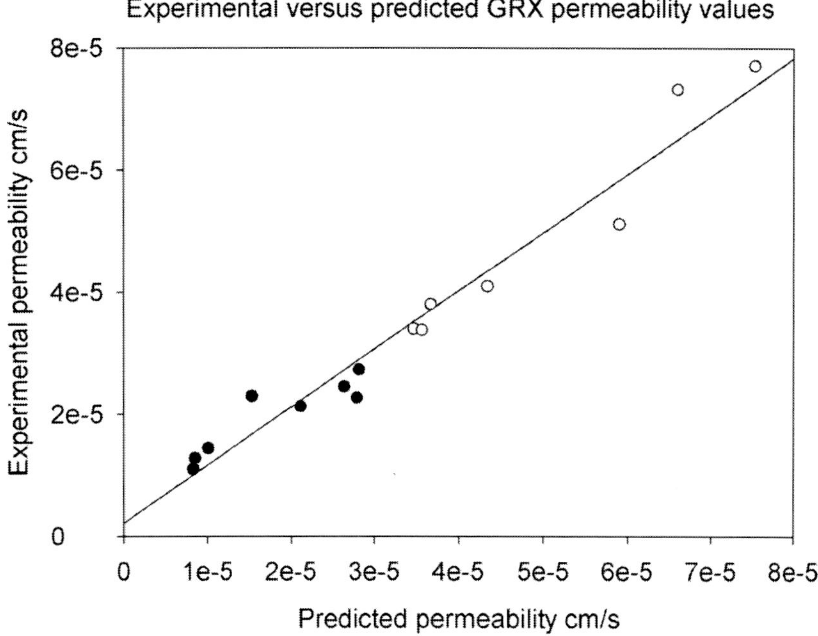

Figure 12. Experimental versus predicted concentrations of grepafloxacin values obtained in rat and in CACO-2 cells.

In the case of grepafloxacin the surface factor value was 4.30 and the expression factor approached 1 and was removed from the model. The difference with respect CNV97100 is that Grepafloxacin is substrate of MRP2 and it could be possible that the expression level of this transporter in our CACO-2 system was similar to the expression in Wistar rat so the correction for the expression of the transporter was not necessary. This mathematical result can be used to establish a rational hypothesis about expression level differences that need to be confirmed by other techniques.

An example of scaling up values from in vitro to human by a model approach is described by Shirasaka Y et al. (Shirasaka et al., 2008b). The authors in first place characterize the relationship between Km and Vmax and expression level by using in vitro models with different expression level and then estimate the Vmax and Km for humans based on human expression level of the transporter. They validated this procedure by measuring rat intestinal permeability and comparing with the simulated values based on predicted Km and Vmax in rat from rat expression level.(Shirasaka et al., 2008a). The controversial point of this work was that the authors modeled the Km value as a function of the expression level while from a theoretical point of view it should depend only on the affinity by the substrate. More recently Tachibana T et al (Tachibana et al., 2010) using a three compartment model (as the one shown previously by Gonzalez-Alvaret I. et al) was able to simulate and reproduce Shirasaka Y et al data with a constant Km value and varying only Vmax as a function of the transporter expression demonstrating how a more accurate and realistic model was able to eliminate the bias in Km estimation.

As a conclusion, these examples show us how mathematical modeling and molecular biology methods are complementary tools in Biopharmaceutics. The new molecular knowledge can help us to develop best predictive models while on the other hand modeling transport and absorption processes could help us in deciphering the molecular mechanisms by kinetic model selection procedures. In addition, the adequate mathematical modeling can help to maximize the information that is obtained from the experiments.

CONCLUSIONS

- CACO-2 cell monolayers are a versatile and useful model for screening drug candidates and with the adequate experimental design are useful to study drug transport mechanisms.
- A detailed description of the data treatment for calculating permeability should be always provided in order to ensure the accuracy of the parameter and its usefulness for quantitative structure-property correlations.
- Mathematical modeling methods are very useful to maximize the information obtained from permeability experiments and are complementary tools to molecular biology approaches to explore transporters involvement in drug permeation.
- In order to scale up the parameters values from in vitro to animals or human models a good modeling of the physiological differences is critical to optimize the predictability of the in vitro models.

REFERENCES

Adachi, Y., Suzuki, H., Sugiyama, Y., (2001). "Comparative studies on in vitro methods for evaluating in vivo function of MDR1 P-glycoprotein". *Pharm Res* 18: 1660-1668.

Adson, A., Raub, T.J., Burton, P.S., Barsuhn, C.L., Hilgers, A.R., Audus, K.L., Ho, N.F., (1994). "Quantitative approaches to delineate paracellular diffusion in cultured epithelial cell monolayers". *J Pharm Sci* 83: 1529-1536.

Amidon, G.L., Lennernas, H., Shah, V.P., Crison, J.R., (1995). "A theoretical basis for a biopharmaceutic drug classification: the correlation of in vitro drug product dissolution and in vivo bioavailability". *Pharm Res* 12: 413-420.

Anderle, P., Rakhmanova, V., Woodford, K., Zerangue, N., Sadee, W., (2003). "Messenger RNA expression of transporter and ion channel genes in undifferentiated and differentiated CACO-2 cells compared to human intestines". *Pharm Res* 20: 3-15.

Artursson, P., (1990). "Epithelial transport of drugs in cell culture. I: A model for studying the passive diffusion of drugs over intestinal absorptive (CACO-2) cells". *J Pharm Sci* 79: 476-482.

Artursson, P., (1991). "Cell cultures as models for drug absorption across the intestinal mucosa". *Crit Rev Ther Drug Carrier Syst* 8: 305-330.

Artursson, P., Karlsson, J., (1991). "Correlation between oral drug absorption in humans and apparent drug permeability coefficients in human intestinal epithelial (CACO-2) cells". *Biochem Biophys Res Commun* 175: 880-885.

Artursson, P., Borchardt, R.T., (1997). "Intestinal drug absorption and metabolism in cell cultures: CACO-2 and beyond". *Pharm Res* 14: 1655-1658.

Artursson, P., Palm, K., Luthman, K., (2001). "CACO-2 monolayers in experimental and theoretical predictions of drug transport". *Adv Drug Deliv Rev* 46: 27-43.

Augustijns, P., Mols, R., (2004). "HPLC with programmed wavelength fluorescence detection for the simultaneous determination of marker compounds of integrity and P-gp functionality in the CACO-2 intestinal absorption model". *Journal of pharmaceutical and biomedical analysis* 34: 971-978.

Avdeef, A., (2010). "Leakiness and size exclusion of paracellular channels in cultured epithelial cell monolayers-interlaboratory comparison". *Pharm Res* 27: 480-489.

Avdeef, A., Tam, K.Y., (2010). "How well can the CACO-2/Madin-Darby canine kidney models predict effective human jejunal permeability?". *J Med Chem* 53: 3566-3584.

Balakrishnan, A., Hussainzada, N., Gonzalez, P., Bermejo, M., Swaan, P.W., Polli, J.E., (2007). "Bias in estimation of transporter kinetic parameters from overexpression systems: Interplay of transporter expression level and substrate affinity". *J Pharmacol Exp Ther* 320: 133-144.

Balimane, P.V., Chong, S., Morrison, R.A., (2000). "Current methodologies used for evaluation of intestinal permeability and absorption". *J Pharmacol Toxicol Methods* 44: 301-312.

Bermejo, M., Avdeef, A., Ruiz, A., Nalda, R., Ruell, J.A., Tsinman, O., Gonzalez, I., Fernandez, C., Sanchez, G., Garrigues, T.M., Merino, V., (2004). "PAMPA--a drug absorption in vitro model 7. Comparing rat in situ, CACO-2, and PAMPA permeability of fluoroquinolones". *Eur J Pharm Sci* 21: 429-441.

Bleasby, K., Chauhan, S., Brown, C.D., (2000). "Characterization of MPP+ secretion across human intestinal CACO-2 cell monolayers: role of P-glycoprotein and a novel Na(+)-dependent organic cation transport mechanism". *Br J Pharmacol* 129: 619-625.

Bohets, H., Annaert, P., Mannens, G., Van Beijsterveldt, L., Anciaux, K., Verboven, P., Meuldermans, W., Lavrijsen, K., (2001). "Strategies for absorption screening in drug discovery and development". *Current topics in medicinal chemistry* 1: 367-383.

Borst, P., Evers, R., Kool, M., Wijnholds, J., (1999). "The multidrug resistance protein family". *Biochim Biophys Acta* 1461: 347-357.

Carriere, V., Lesuffleur, T., Barbat, A., Rousset, M., Dussaulx, E., Costet, P., de Waziers, I., Beaune, P., Zweibaum, A., (1994). "Expression of cytochrome P-450 3A in HT29-MTX cells and CACO-2 clone TC7". *FEBS letters* 355: 247-250.

Collett, A., Sims, E., Walker, D., He, Y.L., Ayrton, J., Rowland, M., Warhurst, G., (1996). "Comparison of HT29-18-C1 and CACO-2 cell lines as models for studying intestinal paracellular drug absorption". *Pharm Res* 13: 216-221.

Conradi, R.A., Wilkinson, K.F., Rush, B.D., Hilgers, A.R., Ruwart, M.J., Burton, P.S., (1993). "In vitro/in vivo models for peptide oral absorption: comparison of CACO-2 cell permeability with rat intestinal absorption of renin inhibitory peptides". *Pharm Res* 10: 1790-1792.

Corti, G., Maestrelli, F., Cirri, M., Zerrouk, N., Mura, P., (2006). "Development and evaluation of an in vitro method for prediction of human drug absorption II. Demonstration of the method suitability". *Eur J Pharm Sci* 27: 354-362.

Cousein, E., Barthelemy, C., Poullain, S., Simon, N., Lestavel, S., Williame, V., Joiris, E., Danel, C., Clavey, V., Brossard, D., Robert, H., Crauste-Manciet, S., Vaccher, C., Odou, P., (2007). "P-glycoprotein and cytochrome P450 3A4 involvement in risperidone transport using an in vitro CACO-2/TC7 model and an in vivo model". *Prog Neuropsychopharmacol Biol Psychiatry* 31: 878-886.

Cho, M.J., Thompson, D.P., Cramer, C.T., Vidmar, T.J., Scieszka, J.F., (1989). "The Madin Darby canine kidney (MDCK) epithelial cell monolayer as a model cellular transport barrier". *Pharm Res* 6: 71-77.

Dahan, A., Amidon, G.L., (2009). "Segmental dependent transport of low permeability compounds along the small intestine due to P-glycoprotein: the role of efflux transport in the oral absorption of BCS class III drugs". *Mol Pharm* 6: 19-28.

Dahan, A., Sabit, H., Amidon, G.L., (2009). "Multiple efflux pumps are involved in the transepithelial transport of colchicine: combined effect of p-glycoprotein and multidrug resistance-associated protein 2 leads to decreased intestinal absorption throughout the entire small intestine". *Drug metabolism and disposition: the biological fate of chemicals* 37: 2028-2036.

Delie, F., Rubas, W., (1997). "A human colonic cell line sharing similarities with enterocytes as a model to examine oral absorption: advantages and limitations of the CACO-2 model". *Crit Rev Ther Drug Carrier Syst* 14: 221-286.

des Rieux, A., Ragnarsson, E.G., Gullberg, E., Preat, V., Schneider, Y.J., Artursson, P., (2005). "Transport of nanoparticles across an in vitro model of the human intestinal follicle associated epithelium". *Eur J Pharm Sci* 25: 455-465.

Engman, H.A., Lennernas, H., Taipalensuu, J., Otter, C., Leidvik, B., Artursson, P., (2001). "CYP3A4, CYP3A5, and MDR1 in human small and large intestinal cell lines suitable for drug transport studies". *J Pharm Sci* 90: 1736-1751.

Evers, R., Zaman, G.J., van Deemter, L., Jansen, H., Calafat, J., Oomen, L.C., Oude Elferink, R.P., Borst, P., Schinkel, A.H., (1996). "Basolateral localization and export activity of the human multidrug resistance-associated protein in polarized pig kidney cells". *J Clin Invest* 97: 1211-1218.

Fernandez-Teruel, C., Gonzalez-Alvarez, I., Casabo, V.G., Ruiz-Garcia, A., Bermejo, M., (2005). "Kinetic modelling of the intestinal transport of sarafloxacin. Studies in situ in rat and in vitro in CACO-2 cells". *Journal of drug targeting* 13: 199-212.

Fine, K.D., Santa Ana, C.A., Porter, J.L., Fordtran, J.S., (1995). "Effect of changing intestinal flow rate on a measurement of intestinal permeability". *Gastroenterology* 108: 983-989.

Flanagan, S.D., Takahashi, L.H., Liu, X., Benet, L.Z., (2002). "Contributions of saturable active secretion, passive transcellular, and paracellular diffusion to the overall transport of furosemide across adenocarcinoma (CACO-2) cells". *J Pharm Sci* 91: 1169-1177.

Fliszar, K.A., Hill, B.T., Foster, N., (2007). "Predicting human drug pharmacokinetics from in vitro permeability using an absorption-disposition model". *J Pharm Sci* 96: 2161-2170.

Gan, L.S., Yanni, S., Thakker, D.R., (1998). "Modulation of the tight junctions of the CACO-2 cell monolayers by H2-antagonists". *Pharm Res* 15: 53-57.

Gonzalez-Alvarez, I., Fernandez-Teruel, C., Garrigues, T.M., Casabo, V.G., Ruiz-Garcia, A., Bermejo, M., (2005). "Kinetic modelling of passive transport and active efflux of a fluoroquinolone across CACO-2 cells using a compartmental approach in NONMEM". *Xenobiotica* 35: 1067-1088.

Gonzalez-Alvarez, I., Fernandez-Teruel, C., Casabo-Alos, V.G., Garrigues, T.M., Polli, J.E., Ruiz-Garcia, A., Bermejo, M., (2007). "In situ kinetic modelling of intestinal efflux in rats: functional characterization of segmental differences and correlation with in vitro results". *Biopharmaceutics & drug disposition* 28: 229-239.

Gonzalez-Alvarez, I., Gonzalez-Alvarez, M., Oltra-Noguera, D., Merino, V., Tokesi, N., Ovadi, J., Bermejo, M., (2009). "Unique pharmacology of KAR-2, a potential anti-cancer agent: absorption modelling and selective mitotic spindle targeting". *Eur J Pharm Sci* 36: 11-19.

Gres, M.C., Julian, B., Bourrie, M., Meunier, V., Roques, C., Berger, M., Boulenc, X., Berger, Y., Fabre, G., (1998). "Correlation between oral drug absorption in humans, and apparent drug permeability in TC-7 cells, a human epithelial intestinal cell line: comparison with the parental CACO-2 cell line". *Pharm Res* 15: 726-733.

He, Y.L., Murby, S., Warhurst, G., Gifford, L., Walker, D., Ayrton, J., Eastmond, R., Rowland, M., (1998). "Species differences in size discrimination in the paracellular pathway reflected by oral bioavailability of poly(ethylene glycol) and D-peptides". *J Pharm Sci* 87: 626-633.

Hidalgo, I.J., Raub, T.J., Borchardt, R.T., (1989). "Characterization of the human colon carcinoma cell line (CACO-2) as a model system for intestinal epithelial permeability". *Gastroenterology* 96: 736-749.

Hu, M., Sinko, P.J., deMeere, A.L., Johnson, D.A., Amidon, G.L., (1988). "Membrane permeability parameters for some amino acids and beta-lactam antibiotics: application of the boundary layer approach". *J Theor Biol* 131: 107-114.

Hubatsch, I., Ragnarsson, E.G., Artursson, P., (2007). "Determination of drug permeability and prediction of drug absorption in CACO-2 monolayers". *Nat Protoc* 2: 2111-2119.

Ingels, F., Beck, B., Oth, M., Augustijns, P., (2004). "Effect of simulated intestinal fluid on drug permeability estimation across CACO-2 monolayers". *Int J Pharm* 274: 221-232.

Irvine, J.D., Takahashi, L., Lockhart, K., Cheong, J., Tolan, J.W., Selick, H.E., Grove, J.R., (1999). "MDCK (Madin-Darby canine kidney) cells: A tool for membrane permeability screening". *J Pharm Sci* 88: 28-33.

Jumarie, C., Malo, C., (1991). "CACO-2 cells cultured in serum-free medium as a model for the study of enterocytic differentiation in vitro". *J Cell Physiol* 149: 24-33.

Katsura, T., Inui, K., (2003). "Intestinal absorption of drugs mediated by drug transporters: mechanisms and regulation". *Drug Metab Pharmacokinet* 18: 1-15.

Kim, D.C., Burton, P.S., Borchardt, R.T., (1993). "A correlation between the permeability characteristics of a series of peptides using an in vitro cell culture model (CACO-2) and those using an in situ perfused rat ileum model of the intestinal mucosa". *Pharm Res* 10: 1710-1714.

Knipp, G.T., Ho, N.F., Barsuhn, C.L., Borchardt, R.T., (1997). "Paracellular diffusion in CACO-2 cell monolayers: effect of perturbation on the transport of hydrophilic compounds that vary in charge and size". *J Pharm Sci* 86: 1105-1110.

Le Ferrec, E., Chesne, C., Artusson, P., Brayden, D., Fabre, G., Gires, P., Guillou, F., Rousset, M., Rubas, W., Scarino, M.L., (2001). "In vitro models of the intestinal barrier. The report and recommendations of ECVAM Workshop 46. European Centre for the Validation of Alternative methods". *Altern Lab Anim* 29: 649-668.

Lee, K., Thakker, D.R., (1999). "Saturable transport of H2-antagonists ranitidine and famotidine across CACO-2 cell monolayers". *J Pharm Sci* 88: 680-687.

Lennernas, H., Nilsson, D., Aquilonius, S.M., Ahrenstedt, O., Knutson, L., Paalzow, L.K., (1993). "The effect of L-leucine on the absorption of levodopa, studied by regional jejunal perfusion in man". *Br J Clin Pharmacol* 35: 243-250.

Lennernas, H., Regardh, C.G., (1993). "Evidence for an interaction between the beta-blocker pafenolol and bile salts in the intestinal lumen of the rat leading to dose-dependent oral absorption and double peaks in the plasma concentration-time profile". *Pharm Res* 10: 879-883.

Lennernas, H., Lee, I.D., Fagerholm, U., Amidon, G.L., (1997). "A residence-time distribution analysis of the hydrodynamics within the intestine in man during a regional single-pass perfusion with Loc-I-Gut: in-vivo permeability estimation". *J Pharm Pharmacol* 49: 682-686.

Lennernas, H., (1998). "Human intestinal permeability". *J Pharm Sci* 87: 403-410.

Lentz, K.A., Hayashi, J., Lucisano, L.J., Polli, J.E., (2000). "Development of a more rapid, reduced serum culture system for CACO-2 monolayers and application to the biopharmaceutics classification system". *Int J Pharm* 200: 41-51.

Lesuffleur, T., Barbat, A., Dussaulx, E., Zweibaum, A., (1990). "Growth adaptation to methotrexate of HT-29 human colon carcinoma cells is associated with their ability to differentiate into columnar absorptive and mucus-secreting cells". *Cancer Res* 50: 6334-6343.

Linnankoski, J., Ranta, V.P., Yliperttula, M., Urtti, A., (2008). "Passive oral drug absorption can be predicted more reliably by experimental than computational models--fact or myth". *Eur J Pharm Sci* 34: 129-139.

Loscher, W., Potschka, H., (2005). "Drug resistance in brain diseases and the role of drug efflux transporters". *Nature reviews* 6: 591-602.

Ma, T.Y., Hollander, D., Bhalla, D., Nguyen, H., Krugliak, P., (1992). "IEC-18, a nontransformed small intestinal cell line for studying epithelial permeability". *J Lab Clin Med* 120: 329-341.

Manach, C., Donovan, J.L., (2004). "Pharmacokinetics and metabolism of dietary flavonoids in humans". *Free radical research* 38: 771-785.

Matsson, P., Bergstrom, C.A., Nagahara, N., Tavelin, S., Norinder, U., Artursson, P., (2005). "Exploring the role of different drug transport routes in permeability screening". *J Med Chem* 48: 604-613.

Miller, J.H., Heath, L.N., (1989). "Growth, enzyme activity, sugar transport, and hormone supplement responses in cells cloned from a pig kidney cell line LLC-PK1". *J Cell Physiol* 139: 538-549.

Miret, S., Abrahamse, L., de Groene, E.M., (2004). "Comparison of in vitro models for the prediction of compound absorption across the human intestinal mucosa". *J Biomol Screen* 9: 598-606.

Nigsch, F., Klaffke, W., Miret, S., (2007). "In vitro models for processes involved in intestinal absorption". *Expert opinion on drug metabolism & toxicology* 3: 545-556.

Nomoto, M., Tatebayashi, T., Morita, J., Suzuki, H., Aizawa, K., Kurosawa, T., Komiya, I., (2009). "Physiological models are good tools to predict rat bioavailability of EF5154 prodrugs from in vitro intestinal parameters". *J Pharm Sci* 98: 1532-1544.

Pade, V., Stavchansky, S., (1997). "Estimation of the relative contribution of the transcellular and paracellular pathway to the transport of passively absorbed drugs in the CACO-2 cell culture model". *Pharm Res* 14: 1210-1215.

Paul, E.C., Hochman, J., Quaroni, A., (1993). "Conditionally immortalized intestinal epithelial cells: novel approach for study of differentiated enterocytes". *Am J Physiol* 265: C266-278.

Pauletti, G.M., Okumu, F.W., Borchardt, R.T., (1997). "Effect of size and charge on the passive diffusion of peptides across CACO-2 cell monolayers via the paracellular pathway". *Pharm Res* 14: 164-168.

Pfrunder, A., Gutmann, H., Beglinger, C., Drewe, J., (2003). "Gene expression of CYP3A4, ABC-transporters (MDR1 and MRP1-MRP5) and hPXR in three different human colon carcinoma cell lines". *J Pharm Pharmacol* 55: 59-66.

Pontier, C., Pachot, J., Botham, R., Lenfant, B., Arnaud, P., (2001). "HT29-MTX and CACO-2/TC7 monolayers as predictive models for human intestinal absorption: role of the mucus layer". *J Pharm Sci* 90: 1608-1619.

Raeissi, S.D., Hidalgo, I.J., Segura-Aguilar, J., Artursson, P., (1999). "Interplay between CYP3A-mediated metabolism and polarized efflux of terfenadine and its metabolites in intestinal epithelial CACO-2 (TC7) cell monolayers". *Pharm Res* 16: 625-632.

Rodriguez-Ibanez, M., Nalda-Molina, R., Montalar-Montero, M., Bermejo, M.V., Merino, V., Garrigues, T.M., (2003). "Transintestinal secretion of ciprofloxacin, grepafloxacin and sparfloxacin: in vitro and in situ inhibition studies". *Eur J Pharm Biopharm* 55: 241-246.

Rodriguez-Ibanez, M., Sanchez-Castano, G., Montalar-Montero, M., Garrigues, T.M., Bermejo, M., Merino, V., (2006). "Mathematical modelling of in situ and in vitro efflux of ciprofloxacin and grepafloxacin". *Int J Pharm* 307: 33-41.

Rubas, W., Cromwell, M.E., Shahrokh, Z., Villagran, J., Nguyen, T.N., Wellton, M., Nguyen, T.H., Mrsny, R.J., (1996). "Flux measurements across CACO-2 monolayers may predict transport in human large intestinal tissue". *J Pharm Sci* 85: 165-169.

Sanchez-Castano, G., Ruiz-Garcia, A., Banon, N., Bermejo, M., Merino, V., Freixas, J., Garriguesx, T.M., Pla-Delfina, J.M., (2000). "Intrinsic absolute bioavailability prediction in rats based on in situ absorption rate constants and/or in vitro partition coefficients: 6-fluoroquinolones". *J Pharm Sci* 89: 1395-1403.

Seithel, A., Karlsson, J., Hilgendorf, C., Bjorquist, A., Ungell, A.L., (2006). "Variability in mRNA expression of ABC- and SLC-transporters in human intestinal cells: comparison between human segments and CACO-2 cells". *Eur J Pharm Sci* 28: 291-299.

Shimizu, R., Sukegawa, T., Tsuda, Y., Itoh, T., (2008). "Quantitative prediction of oral absorption of PEPT1 substrates based on in vitro uptake into CACO-2 cells". *Int J Pharm* 354: 104-110.

Shirasaka, Y., Masaoka, Y., Kataoka, M., Sakuma, S., Yamashita, S., (2008a). "Scaling of in vitro membrane permeability to predict P-glycoprotein-mediated drug absorption in vivo". *Drug metabolism and disposition: the biological fate of chemicals* 36: 916-922.

Shirasaka, Y., Sakane, T., Yamashita, S., (2008b). "Effect of P-glycoprotein expression levels on the concentration-dependent permeability of drugs to the cell membrane". *J Pharm Sci* 97: 553-565.

Skolnik, S., Lin, X., Wang, J., Chen, X.H., He, T., Zhang, B., (2010). "Towards prediction of in vivo intestinal absorption using a 96-well CACO-2 assay". *J Pharm Sci* 99: 3246-3265.

Stewart, B.H., Chan, O.H., Lu, R.H., Reyner, E.L., Schmid, H.L., Hamilton, H.W., Steinbaugh, B.A., Taylor, M.D., (1995). "Comparison of intestinal permeabilities determined in multiple in vitro and in situ models: relationship to absorption in humans". *Pharm Res* 12: 693-699.

Sugano, K., Kansy, M., Artursson, P., Avdeef, A., Bendels, S., Di, L., Ecker, G.F., Faller, B., Fischer, H., Gerebtzoff, G., Lennernaes, H., Senner, F., (2010). "Coexistence of passive and carrier-mediated processes in drug transport". *Nat Rev Drug Discov* 9: 597-614.

Sun, D., Lennernas, H., Welage, L.S., Barnett, J.L., Landowski, C.P., Foster, D., Fleisher, D., Lee, K.D., Amidon, G.L., (2002). "Comparison of human duodenum and CACO-2 gene expression profiles for 12,000 gene sequences tags and correlation with permeability of 26 drugs". *Pharm Res* 19: 1400-1416.

Tachibana, T., Kitamura, S., Kato, M., Mitsui, T., Shirasaka, Y., Yamashita, S., Sugiyama, Y., (2010). "Model analysis of the concentration-dependent permeability of P-gp substrates". *Pharm Res* 27: 442-446.

Taipalensuu, J., Tornblom, H., Lindberg, G., Einarsson, C., Sjoqvist, F., Melhus, H., Garberg, P., Sjostrom, B., Lundgren, B., Artursson, P., (2001). "Correlation of gene expression of ten drug efflux proteins of the ATP-binding cassette transporter family in normal human jejunum and in human intestinal epithelial CACO-2 cell monolayers". *J Pharmacol Exp Ther* 299: 164-170.

Tanigawara, Y., Okamura, N., Hirai, M., Yasuhara, M., Ueda, K., Kioka, N., Komano, T., Hori, R., (1992). "Transport of digoxin by human P-glycoprotein expressed in a porcine kidney epithelial cell line (LLC-PK1)". *J Pharmacol Exp Ther* 263: 840-845.

Tavelin, S., Milovic, V., Ocklind, G., Olsson, S., Artursson, P., (1999). "A conditionally immortalized epithelial cell line for studies of intestinal drug transport". *J Pharmacol Exp Ther* 290: 1212-1221.

Tavelin, S., Grasjo, J., Taipalensuu, J., Ocklind, G., Artursson, P., (2002). "Applications of epithelial cell culture in studies of drug transport". *Methods Mol Biol* 188: 233-272.

Tavelin, S., Hashimoto, K., Malkinson, J., Lazorova, L., Toth, I., Artursson, P., (2003a). "A new principle for tight junction modulation based on occludin peptides". *Mol Pharmacol* 64: 1530-1540.

Tavelin, S., Taipalensuu, J., Soderberg, L., Morrison, R., Chong, S., Artursson, P., (2003b). "Prediction of the oral absorption of low-permeability drugs using small intestine-like 2/4/A1 cell monolayers". *Pharm Res* 20: 397-405.

Terada, T., Saito, H., Mukai, M., Inui, K., (1997). "Characterization of stably transfected kidney epithelial cell line expressing rat H+/peptide cotransporter PEPT1: localization of PEPT1 and transport of beta-lactam antibiotics". *J Pharmacol Exp Ther* 281: 1415-1421.

Turco, L., Catone, T., Caloni, F., Consiglio, E.D., Testai, E., Stammati, A., (2010). "CACO-2/TC7 cell line characterization for intestinal absorption: How reliable is this in vitro model for the prediction of the oral dose fraction absorbed in human?". *Toxicol In Vitro.*

Vachon, P.H., Beaulieu, J.F., (1992). "Transient mosaic patterns of morphological and functional differentiation in the CACO-2 cell line". *Gastroenterology* 103: 414-423.

Volpe, D.A., (2010). "Application of Method Suitability for Drug Permeability Classification". *AAPS J.*

Walter, E., Kissel, T., (1994). "Transepithelial transport and metabolism of thyrotropin-releasing hormone (TRH) in monolayers of a human intestinal cell line (CACO-2): evidence for an active transport component?". *Pharm Res* 11: 1575-1580.

Walter, E., Kissel, T., Reers, M., Dickneite, G., Hoffmann, D., Stuber, W., (1995). "Transepithelial transport properties of peptidomimetic thrombin inhibitors in monolayers of a human intestinal cell line (CACO-2) and their correlation to in vivo data". *Pharm Res* 12: 360-365.

Walter, E., Janich, S., Roessler, B.J., Hilfinger, J.M., Amidon, G.L., (1996). "HT29-MTX/CACO-2 cocultures as an in vitro model for the intestinal epithelium: in vitro-in vivo correlation with permeability data from rats and humans". *J Pharm Sci* 85: 1070-1076.

Watanabe, T., Onuki, R., Yamashita, S., Taira, K., Sugiyama, Y., (2005). "Construction of a functional transporter analysis system using MDR1 knockdown CACO-2 cells". *Pharm Res* 22: 1287-1293.

Wikman, A., Karlsson, J., Carlstedt, I., Artursson, P., (1993). "A drug absorption model based on the mucus layer producing human intestinal goblet cell line HT29-H". *Pharm Res* 10: 843-852.

Wils, P., Warnery, A., Phung-Ba, V., Scherman, D., (1994). "Differentiated intestinal epithelial cell lines as in vitro models for predicting the intestinal absorption of drugs". *Cell Biol Toxicol* 10: 393-397.

Wilson, F.A., Dietschy, J.M., (1974). "The intestinal unstirred layer: its surface area and effect on active transport kinetics". *Biochim Biophys Acta* 363: 112-126.

Yamaguchi, T., Ikeda, C., Sekine, Y., (1986). "Intestinal absorption of a beta-adrenergic blocking agents nadolol. II. Mechanism of the inhibitory effect on the intestinal absorption of nadolol by sodium cholate in rats". *Chem Pharm Bull (Tokyo)* 34: 3836-3843.

Yamashita, S., Furubayashi, T., Kataoka, M., Sakane, T., Sezaki, H., Tokuda, H., (2000). "Optimized conditions for prediction of intestinal drug permeability using CACO-2 cells". *Eur J Pharm Sci* 10: 195-204.

Zhang, W., Lim, L.Y., (2008). "Effects of spice constituents on P-glycoprotein-mediated transport and CYP3A4-mediated metabolism in vitro". *Drug metabolism and disposition: the biological fate of chemicals* 36: 1283-1290.

Zhang, X., Shedden, K., Rosania, G.R., (2006). "A cell-based molecular transport simulator for pharmacokinetic prediction and cheminformatic exploration". *Mol Pharm* 3: 704-716.

In: CACO-2 Cells and their Uses
Editor: Megan A. Schulz, pp. 165-180
ISBN 978-1-61209-917-0

Chapter 10

NANOCARRIER ABSORPTION STUDIES WITH CACO-2 CELLS

Frederic Lagarce[1,2*] ***and Emilie Roger***[3]
[1]Inserm U646, Ingéniérie de la vectorisation particulaire, Angers University, France
[2]Pharmaceutical department, Angers Universitary Hospital, France
[3]Laboratoire de pharmacie Biogalénique, Strasbourg university, France

ABSTRACT

Caco-2 cells are widely and commonly used to study oral absorption of drugs. Colloidal drug delivery systems, such as nanoparticles, nanocapsules, liposomes, micelles, have been proposed to enhance the absorption of poorly absorbed and poorly soluble drugs (BCS class IV). Caco-2 cells can be used to study, compare and screen the performances of these drug delivery systems. In this chapter, experimental studies with nanocarriers and Caco-2 cells will be described and analyzed. Two different types of studies will be considered: absorption studies (apparent permeability determination though cell monolayers under various experimental conditions) and mechanistic studies (determination of the main routes of uptake by the cells and the ratio between passive and active pathways). For each type, examples will be given and analyzed in order to shed light on the envisioned findings obtained with Caco-2 cell tool but also on the pitfalls of these studies. In a last part of the chapter, Caco-2 cell monolayers, will be compared with other cellular models and biopharmaceutical tools such as PAMPA's or Single pass perfusion model.

INTRODUCTION

Nanotechnology is a field of growing interest. It has been studied in the area of medicine to enhance the performances of the drugs encapsulated in nanostructures, in order to optimize their distributions in the organism after administration and finally to help them reach their

* Corresponding author : e-mail : frederic.lagarce@univ-angers.fr

pharmacological targets. Many different nano-objects belong to the nanocarrier's family. Nanocarriers are therapeutic formulations with a size under 1μm (often under 200nm). Liposomes, which are formed by one or several phospholipid bilayers mimicking the cell membrane, have been developed more than forty years ago. Nanoparticles can be made of polymer or other components such as lipids. It is possible to obtain a core/shell structure, in this case the nanoparticles are called nanocapsules other way they can be called nanospheres. Nanot echnologies produce many different therapeutic objects of the nanometer range besides liposomes and nanoparticles. Dendrimers, polymeric micelles, nanoemulsion, nanotubes, quantum dots have been also described. In most cases these nanocarriers are intended for the intravenous route of administration but their particular properties have been used to enhance the bioavailability of drug via the oral route. In fact, bioavailability is very often impaired by a poor absorption process. The absorption process, called the biopharmaceutic phase, is closely related to the solubility of the drug and its permeability across the intestinal barrier and these two parameters can be greatly enhanced by the encapsulation of the drug in nanostructures. In order to obtain the proof of concept that nanocarrier can be used for oral delivery, new formulation have been studied on *in vitro* and *ex vivo* models to evaluate their properties to enhance oral absorption. One of the most widely used cellular models to screen formulations is CACO-2 cell monolayers. Indeed, this system is easy to implement and relatively fast (in comparison to *in vivo* full pharmacokinetic studies). But, is it possible to use CACO-2 to study nanocarriers absorption like with any drug? Are there most suitable models? What are the advantages and drawbacks of the cell model in comparison to other more simple or more sophisticated ones? In this chapter, we will discuss these important questions. In a first part, methods to study nanocarrier's permeability with CACO-2 cells will be described, then the results obtained will be reviewed taking into account recent published data in the scientific literature. Finally, the limits and the contributions of CACO-2 cell model in comparison to alternative permeability investigating tools will be discussed.

I. METHOD

a. Absorption Study

To study the transport of colloidal drug delivery systems, CACO-2 cells are seeded onto filter. CACO-2 cells are firstly plated on 75cm^2 flask and incubated in a humidified incubator 5% CO2/95% air atmosphere at 37°C. Most often, cells are cultured in Dulbecco's modified Eagle medium (D-MEM, high glucose) supplemented with 10-20% (v/v) foetal bovine serum, 1% (v/v) non-essential amino acids, 1% (v/v) sodium pyruvate, and 1% antibiotic solution (1x10^{4}UI/mL penicillin, 10 mg/mL streptomycin, 25 μg/mL amphothericin B). After having reached 80% of confluence, cells are harvested and seeded onto insert in cell culture chambers (Figure 1), e.g. Transwell® cell culture chambers (Corning Costar, Cambridge, MA). The culture medium is replaced at the apical and basolateral sides every two days for the first two weeks and everyday thereafter. After 21-25 days in culture, CACO-2 cells are differentiated, and form a confluent monolayer with several properties characteristic of differentiated intestinal epithelial cells as microvilli on the apical side, tight junction between cells, expression of small intestinal hydrolase enzymes, expression of transport system [1] and can be used to study absorption of drug. Controls are performed to determine the formation of the monolayer. The most common is the transepithelial electrical resistance (TEER) of the monolayer, which is measured using an epithelial voltohmmeter.

Before absorption study, cell monolayers are washed with medium of dilution (e.g. Hank's buffered salt solution (HBSS)). Indeed, colloidal drug delivery system is preferably diluted: the medium of dilution and the concentration of tested solution depend on the study. Normally, the transport of colloidal drug delivery systems and other compound are studied in the apical to basolateral direction on CACO-2 cells at 37°C. Consequently, the experiment is started by adding test solution at the apical side and medium at the basolateral side. To study the transport across CACO-2 cell monolayers, inserts are incubated at 37°C. After desired times of incubation (depending on the study, mainly 2h), sample are taken from apical and basolateral sides and compound content is determined.

Sensitive analytical method allows analyzing compound content as LC-MS/MS, HPLC coupled with fluorescent detector, or liquid scintillation counting (in the case of radiolabeled compounds).

Finally, apparent permeability coefficient (P_{app}), expressed in centimeters per second is calculated according to the following equation [2, 3]: $\mathbf{P_{app}=(dQ/dt)\times 1/(A.C_0)}$, where dQ/dt is the rate of drug appearance on the basolateral side ($\mu g.s^{-1}$), C_0 is the initial concentration over the apical side ($\mu g.mL^{-1}$) and A is the surface area of monolayer (cm^2).

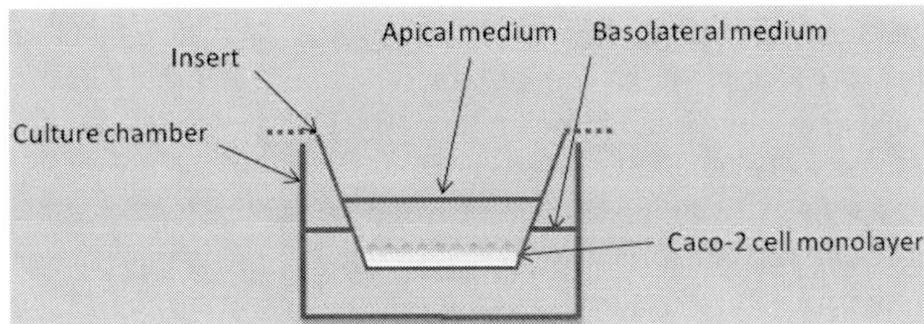

Figure 1. Schematic representation of CACO-2 cell model.

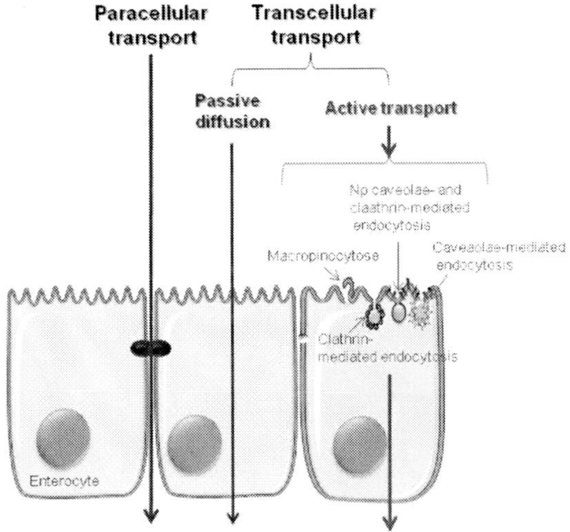

Figure 2. Possible transport mechanism across enterocyte (: lipid raft, : caveolin, : Clathrin, : Tight junction).

b. Mechanistic Studies

Moreover, *in vitro* CACO-2 cell model is used to determine the underlying mechanism responsible for drug delivery transport. Experiments are also performed after differentiation of cells. To cross the intestinal barrier, colloidal system can be transported according to a number of different passive and active transcellular pathways: paracellular or/and transcellular (Figure 2). Different methods are used to determine which mechanism is used by the colloid drug delivery system.

- Study of paracellular transport

Paracellular transport is limited by the presence of tight junction. Consequently, first method to study paracellular transport consists in analyzing the integrity of tight junctions. To control the tightness of junctions, the transepithelial electrical resistance (TEER) of monolayers is measured. A decrease of TEER after incubation with colloidal system in comparison to initial value of TEER, reflects an opening of tight junction, and consequently, a potential paracellular transport [4]. The paracellular pathway can also be assessed by mannitol flux (hydrophilic molecule transport through the tight junction). Generally, the apical to basolateral permeability of mannitol is performed with [^{14}C] mannitol and samples are analyzed by liquid scintillation counting. Typically, P_{app} is higher and more variable (0.5-2 x 10^{6}cm/s); this is related to an opening of tight junction. Other molecules are also used to control paracellular transport: Lucifer Yellow, fluorescein, atenolol… [4, 5]. Expression of tight junction proteins occludin (ZO-1) or claudin-4 by confocal microscopy can also be observed. Because CACO-2 cell model presents some versatility with significant differences between laboratory, it very important to well characterize model and determine values of TEER, or mannitol permeability [1] before performing any experiment.

- Study of transcellular transport

To demonstrate transcellular transport, it is usual to analyze nanoparticles uptake. Different methods can be performed [6]. The use of particle labeled with different fluorochromes allows following cell uptake by confocal microscopy or flow cytometry (FACS as fluorescent activated cell sorter). Radiolabeled compound is also used and intracellular radioactivity is then determined by liquid scintillation counting. Moreover, direct dosage of encapsulated drug by a sensitive method such as LC-MS/MS allows the determination of intracellular drug onto cell monolayers. Finally, electronic microscopy enables to observe particles inside cells.

Then, transcellular mechanisms include passive (diffusion) or active (endocytosis) mechanisms. To distinguish if it is an energy-dependent process or not, absorption study or uptake experiment can be performed at 4°C instead of 37°C to prevent energy process. Another method consists of the inhibition of ATP synthesis with inhibitor (e.g. 2,4-dinitrophenol, sodium azide).

Finally, energy-dependent uptake by enterocytes is described by pinocytosis, which includes macropinocytosis, or endocytosis, which includes clathrine-mediated endocytosis, caveolae-mediated endocytosis and clathrine- and caveolae-independent endocytosis [7] [8].

Study in the presence of specific inhibitor for different types of endocytosis allows the determination of particles uptake mechanism. For example, 5-(*N*-ethyl-*N*-isopropyl) amiloride is used as an inhibitor of macropinocytosis, chlorpromazine, phenylarsine oxide and monodansyl Cadaverine are used as inhibitors of clathrin-mediated endocytosis and filipin and genistein are used as caveolar endocytosis, respectively. Nystatin, methyl-b-cyclodextrin and lovastatin are used as an inhibitor of lipid-rafts mediated endocytosis, including caveolae and macropinocytosis [9-11].

II. RESULTS

a. Absorption Study

Transport study across CACO-2 cell model is widely used to determine *in vitro* intestinal transport of drug. In case of colloidal system, majority of work compared the transport of drug-loaded colloidal system and transport of free drug. So, transport of drug encapsulated in various colloidal drug delivery systems was extensively studied across CACO-2 cells. For example, Zha *et al.* [12] demonstrated that chromium nanoparticles (P_{app}=6.9 ± 1.0 × 10^{-6} cm/s) improved the apparent permeability of chromium in comparison to chromium chloride (P_{app}=1.17 ± 0.20 × 10^{-6} cm/s). Song *et al.* showed that proliposome increased by a factor 3.6 the apparent permeability of salmon calcitonin [13]. And Zhang *et al.*, which worked with polymeric micelles systems, demonstrated that cylcosporin A-loaded micelles enhanced significantly the apparent permeability of cyclosporin A in comparison to free cyclosporin A (3.82 ± 1.09 x 10^{-6} cm/s and 0.78 ± 0.06 x 10^{-6} cm/s, respectively) [14].Thus, an improvement of apparent permeability of drug after encapsulation in colloidal system can occur. A superior intestinal absorption can lead to a bioavailability improvement as demonstrated with some colloidal drug delivery systems (table I).

Table I. Endocytosis inhibitors

Endocytosis mecanisms	Inhibitors	Concentration	Ref.
macropinocytose	5-(*N*-ethyl-*N*-isopropyl) amiloride	15 μM	[10]
lipid rafts mediated endocytosis	Nystatin	50 μM	[10]
	lovastatin + methyl-β-cyclodextrin	1.5 μM+ 10 mmol/L	[9]
Clathrin mediated endocytosis	Phenylarsine Oxide	1 μM	[11]
	Monodansyl Cadaverine	300 μM	[11]
	Chlorpromazine	3 μM	[10]
		0.06 μM	[9]
Caveolin-mediated endocytosis	Filipin	4 μM	[11]
		0.6 μM	[9]
	Genistein	100 μM	[11]
Dynamin-mediated endocytosis	Dynasore	50 μM	[11]

Moreover, one explanation of a low oral bioavailability of drugs such as anticancer agents is their efflux by P-glycoprotein (P-gp), which is extensively present along intestinal tract [15]. In CACO-2 cell model, P-gp is generally over expressed [16, 17]. Consequently, the interest of nanocarrier to circumvent the P-gp efflux and improve drug absorption has to be determined. Indeed, numerous scientifics works have demonstrated the capacity of nanocarriers, more especially of excipients to revert multidrug resistant phenomenon. For example, Solutol®, excipients used in the formulation of lipid nanocapsules have the ability to reverse multidrug resistant [18-20] and the inhibitor activity of lipid nanocapsules in P-gp efflux was also demonstrated [21]. Similarly results were obtained with Peceol® excipient and mixed micelles composed of Peceol® which enhance absorption of amphobericine B [22].

Moreover, absorption study across CACO-2 cell model allows also screening different formulation of colloidal drug delivery or physical properties of a type of formulation. Liu *et al.* [28] studied the capability of three different lipid nanocarrier of N3-*o*-toluyl-fluorouracil (TFu) to cross CACO-2 cell monolayers. It was observed that either cationic ($8.49 \pm 1.6 \times 10^{-6}$ cm/s) or anionic ($7.89 \pm 3.0 \times 10^{-6}$ cm/s) SLN (Solid Lipid Nanoparticles) could significantly increase the transport of TFu compared with liposomes ($5.99 \pm 3.2 \times 10^{-6}$ cm/s). Moreover, Roger *et al.* [9] studied the influence of size of paclitaxel-loaded lipid nanocapsules (LNCs). At 37°C, no significant difference was observed on the transport of paclitaxel across CACO-2 cells between LNCs with a size of 25nm, 50nm, 100nm and 130nm.

Table II. Example of results of transport study across CACO-2 ell model correlated with *in vivo* study

Colloidal system delivery	Encapsulated compound	Absorption study across CACO-2 cell model	Bioavailibility *in vivo* study	Ref.
N-trimethyl chitosan nanoparticles	FITC/Urease	P_{app} is 1.4- and 1.6 fold higher than FITC-BSA	Immune response is higher than urease solution	[23]
Sodium taurocholate Liposome	Insulin	P_{app} is 1.8-fold higher than insulin solution	blood glucose is 1.1-fold higher than insulin solution	[24]
Silybin	Nanosuspension	P_{app} is 3-fold higher than un-milled silybin	Bioavailibility is 2,9-fold higher than un-milled silybin nanosuspension	[25]
Polyethylene glycol-coated liposomes	Recombinant human epidermal growth factor (rhEGF)	P_{app} is 3-fold higher than rhEGF	Bioavailability is 1.7- and 2.5-fold higher than rhEGF	[26]
Lipid nanocapsule	Paclitaxel	P_{app} is 3.5-fold higher than free paclitaxel	Bioavailability is 3-fold higher than Taxol®	[9, 27]

b. Mechanistics Study

As described previously, different methods allow elucidating transport mechanism of colloidal drug delivery system. Some works describe the mechanism of transport of these systems and more especially of their encapsulated compound. Recently no fate of colloidal system after crossing intestinal barrier is elucidated [29].

Firstly, using confocal microscopy (Figure 3) or flow cytometry, some works studied the transport of various colloidal systems by paracellular route. Dendrimers [30-33], polymeric micelles [34] or chitosan nanoparticles [23, 35-37] have the ability to open tight junction that permit paracellular transport Figure 3, Nevertheless, this opening is reversible and barrier recovered its tightness properties when carriers were removed.

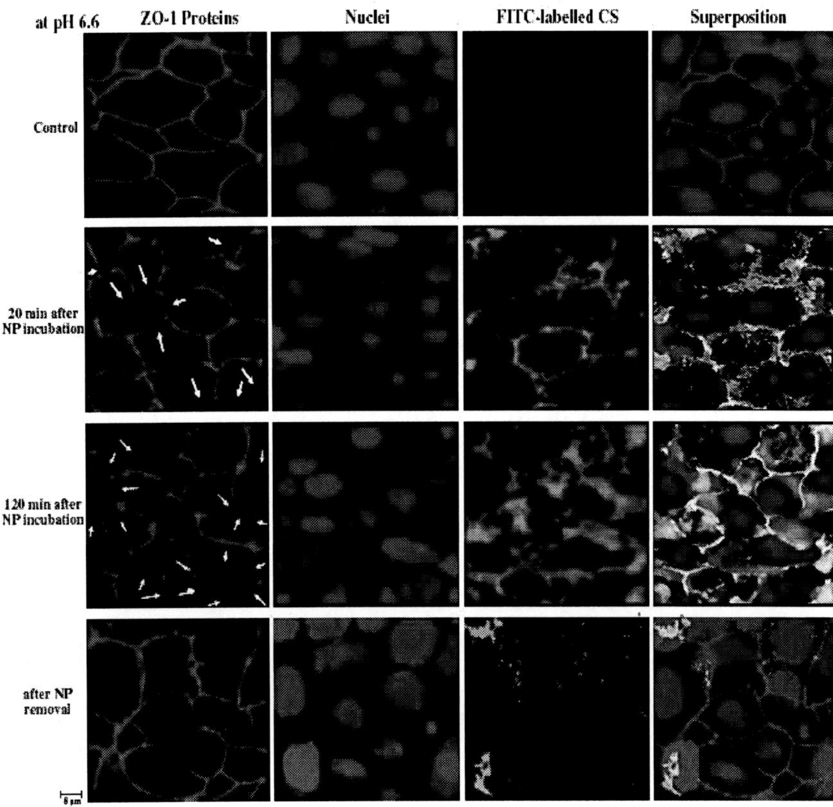

Figure 3. Example of effects of nanoparticles on tight junction structure observed by confocal microscopy. Fluorescence images of CACO-2 cell monolayers immunofluorescently stained for ZO-1 proteins and nuclei after incubation with the FITC-labelled nanoparticles (NP) for 20 or 120 min (the opening of tight junctions was indicated by white arrows) and their counterparts after removal of NPs. Control group: the group without incubation with NPs. CS: chitosan [38].

Figure 3, ZO-1 proteins (protein which composed tight junction) was observed before, during, and after incubation of CACO-2 cell monolayer with nanoparticles. After incubation, some openings of tight junctions were observed but after elimination of nanoparticles, ZO-1 labeling was homogenous. Moreover, opening of tight junction was observed with TEER values and as was shown on figure 4, after removal of these nanoparticles, TEER values that was decreased, recovered initial values. Then, uptake of colloidal system is extensively described with different method, as example, uptake of lipid nanocapsules [9], poly(lactide) nanoparticles [39, 40], polymeric micelles [41] and dendrimers [30] were demonstrated. One extensively quantitative method used consists of measured fluorescence by flow cytometry.

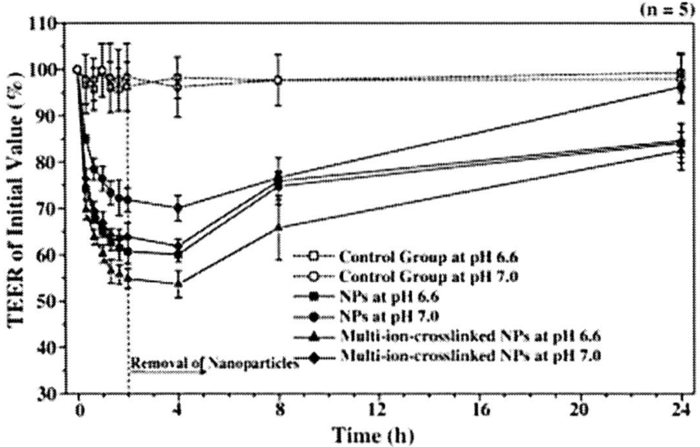

Figure 4. Example of effects of nanoparticles on TEER values of CACO-2 cell monolayers [36].

Other methods widely used are qualitatives: confocal or electronic microscopy. For example, Zhang and Feng [40] or Roger *et al.* [9] showed by confocal microscopy that poly(lactide)-vitamin E TPGS (PLA) nanoparticles or lipid nanocapsules (LNCs) were internalized by CACO-2 cells (Figure 5). Jevprasesphant *et al.* [42] demonstrated uptake and more exactly endocytosis of dendrimer by electronic microscopy.

Specific mechanisms of transport could be also studied using inhibitor agents on CACO-2 cell monolayers. Results demonstrated that the majority of drug encapsulated in carrier were transported by an energy-dependent process (decreased of transport at 4°C) [12, 31, 43]. For example, apparent permeability of paclitaxel-loaded lipid nanocapules with a size of 50nm was $1.35 \pm 0.4 \times 10^6$ cm/s at 4°C and $5.74 \pm 1.21 \times 10^6$ cm/s at 37°C [9].

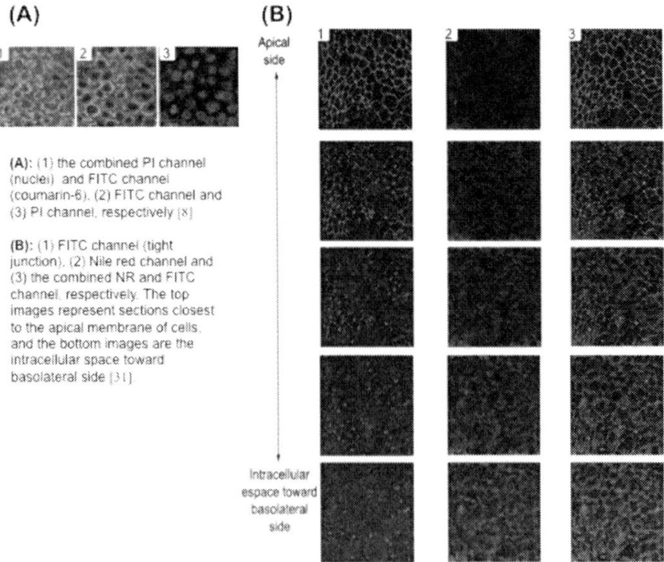

Figure 5. Example of confocal microscopic image of CACO-2 cells after incubation with (1) coumarin-6 loaded PLA-TPGS nanoparticles, (2) Nile red-loaded LNCs [9, 40].

Moreover, in function of colloidal system, active internalization mechanisms by the cells were different. Study in presence of specific inhibitor of each route allows to elucidating it. Like this, internalization by cells is through:

- macropinocytose (e.g. polystyrene nanoparticles: inhibition of transport was 80% in the presence of 5-(*N*-ethyl-*N*-isopropyl) [10] fluid phase endocytosis (e.g. polymeric micelles [44]: transport across CACO-2 cell monolayers with azithromycin decreased the polymer of micelles composition apparent permeability by 34.5%),
- clathrin-dependent pathway (e.g. dendrimer [11, 32, 45, 46] and chitosan nanoparticles: apparent permeability of [^{3}H]-dendrimer was 11.6 x 10^{-6} cm/s and decreased to 7.3 x 10^{-6} cm/s in the presence of filipin),
- caveolae-dependent pathway
- both Clathrin- and caveolae-depend pathways (e.g. lipid nanoparticles [9] and wheat germ agglutinin-conjugated nanoparticles [39] : apparent permeability of paclitaxel-loaded lipidic nanocapsules decreased by a factor 1.2 and 2.8 in the presence of filipin and chlorpromazine, respectively, same inhibitors reduced also the cellular uptake of WGA-nanoparticles)

Furthermore, this model and more especially uptake experiment offers the possibility of comparing system and studying the influence of particle characteristics to pass through intestinal barrier (e.g. type of system, structure, size, nature of surface, charge, zeta potential, and coating as adhesions factors). Thus, Win and Feng demonstrated that uptake of polymeric nanoparticles is size depends. Indeed, efficiency uptake of nanoparticles was above 15% with nanoparticles with a size of 200nm and above 9% with nanoparticle with a size of 1000nm [43]

III. Disadvantage/ Limit of CACO-2 to Study Drug Transport / Other Method

CACO-2 cell monolayers are a model to study drug absorption. As for every model, there are limits in describing the real studied phenomena. In this part, CACO-2 cell models will be compared to other models in order to bring their limits to the foreground. The comparison will focus on nanocarrier permeability studies more details on general comparison between models can be found in chapter 9 of this book. The alternative models will be described from the more artificial to the closest to human's.

a. Parallel Artificial Membrane Permeation Assay (PAMPA)

PAMPA was introduced in 1998 by Kansy [47], the objective was to study passive permeation process. This model is a high-throughput screening tool. It is fastest and simplest to implement in comparison to CACO-2 or other cell-based culture models. In PAMPA, a hydrophobic filter is coated with a mixture of phospholipids and other membrane constituents dissolved in an organic solvent. There are different types of PAMPA's based on the nature of the filter, lipids and transport media used. To our knowledge, only two papers report the use of PAMPA to study permeability of nanocarriers. In the first study by Sandhya *et al.* [48], permeation of liposomal formulation of serratiopeptidase are tested on CACO-2 and using PAMPA. The size of the liposomes is in the range of 78 to 120nm depending on the formulation. The apparent permeability values obtained with the PAMPA show that the liposomes have a potential to enhance the permeation of serratiopeptidase through artificial membrane. Similar results have been found with a CACO-2 model in the same study. In the Sandhya study, the drug is detected but not the liposomes, it is thus impossible to know if the permeability enhancement is due to a transport of the carrier through the membrane or only a better diffusion of the drug associated with liposome compounds. In the second study [49] permeability of exemestane proliposomes is evaluated on CACO-2 cells, *ex vivo* in rat intestine strips and using PAMPA versus pure drug. The authors found that the inclusion in proliposomes enhanced the permeability in every model, but the exemestane permeability did not follow an identical trend in PAMPA, CACO-2 and rat intestine models. Good correlation between PAMPA and CACO-2 indicated that the predominant mechanism for permeation was passive diffusion. The authors attributed the observed differences on permeability measurements between the three models on the permeation mechanisms operating in these models. For example cell tight junctions are expressed in CACO-2 and rat intestine and not in artificial membrane. The comparison of the obtained results with different models may be a solution to understand the permeation mechanisms of the drug and possibly on how the nanocarriers facilitate (or not) the absorption process. PAMPA is not very suitable for studying nanocarriers permeability because in artificial membrane there are no active transport and no possibility for endocytosis. Indeed it has been shown that nano-objects, such as vesicles of nanometer range (nanoparticles, micelles, liposome), are able to enter into the enterocyte via endocytosis (clathrin or caveolae pathways). In PAMPA, the only possible way to cross the membrane is via passive diffusion.

b. Alternative Cell Culture Models

CACO-2 cells is a widely used model because it is relatively simple to implement, but it is not very close to the reality of the absorption in humans. In fact, because of their tumor origin, these cells over express the P-gP. There are other differences from the *in vivo* physiological conditions as the CACO-2 do not produce mucus, which is a major barrier to absorption and there are also no M-like-Cells in this model. Finally the metabolic enzymes (Cytochromes) can display a different activity level from real enterocytes.

In order to make this model more realistic, subclones with a phenotype closer to the human intestinal barrier have been studied and used. For example, the TC7 subclone displaying a P-glycoprotein activity closer to the human enterocyte has been found reliable [50]. Nevertheless, one of the key feature absent from this model is mucus. Mucus, covers the gastro-intestinal epithelium, and is composed of mucin, a large polyssacharide (500,000 Da Subunits), fixed on a protein backbone. Mucus is 95% water and retain particles of size a few micrometers or even less. Mucus is produced by goblet cells in the intestine. The diffusion of nanocarriers in mucus depends on their size and obviously on their surface state (charge, chemical entities, hydrophilic/lipophilic balance). The average thickness of mucus in the gastro intestinal tract is in the range of 10 μm (ileal area) to 170 μm (gastric area) ie over

1000 diameters of a 10 nm particle. It is thus really important to study the diffusion in mucus during the development of new nanocarriers. As CACO-2 do not express mucus, because this model lacks goblet cells, it is essential to use alternative models to take into consideration this characteristic of the absorption process. After oral administration, conventional nanoparticles are trapped in mucus via steric or adhesive forces and eliminated via direct transit through the GI tract [51]. One of the proposed solutions is to use adenocarcinoma HT29 goblet cells. Several clones have been selected after various culture conditions. The clone HT29-H has been widely used and permits to investigate the influence of mucus on absorption, but was phenotypically instable. Another sub-clone, HT29-MTX, selected by differentiation under methotrexate, was found more stable in expressing mucus and can be grown as monolayers [52]. Moreover, HT29 derived models display less expression of tight junction whereas CACO-2 overexpress tight junctions. This has also an impact on the paracellular pathways of drugs. Many papers have described the coculture of CACO-2 cells with HT29 cells thus combining the two major cell types of the intestinal barrier (enterocytes and goblet cells). The conditions for co-culturing have been various and an optimized design was described recently [53]. The study from C. Prego *et al.* [54] shed light on the differences observed with CACO-2 and co-culture models featuring mucus. The authors prepared chitosan nanocapsules encapsulating salmon calcitonin and investigated their uptake on CACO-2 alone or co-cultured with HT29-M6 cells. With fluorescent staining they observed on the co-culture that the chitosan nanocapsules had muco adhesive properties. Results can be very different with both models because of the mucus properties. Thus Dong *et al.* [55], developed bioadhesive poly(D,L-lactide-co-glycolide)/montmorillonite nanoparticles and evaluated their cellular uptake on CACO-2 or HT29 cells after encapsulating a fluorescent 6 coumarin probe. They found a discrepancy between uptake with CACO-2 cells and HT29 cells. In fact the uptake was much lower in HT29 cells in comparison to CACO-2 cells (11% vs 57%) and the enhancement of uptake obtained with addition of montmorillonite was 177% for CACO-2 cells vs 55% for HT29 cells. This result was attributed to a difference in uptake capacity of HT29 cells in comparison with CACO-2 cells, but the effect of mucus cannot be ignored. Very few studies describe the use of alternate cell lines such as HT29 to study the absorption of nanocarriers through cell monolayer. In a few published examples one find either similarities either discrepancies between regular CACO-2 and alternative cell lines on absorption or uptake studies. Even if CACO-2 is a well established model, it does not describe the real physiological phenomena and this is critical with nanocarriers because their size or surface state can impair their diffusion possibilities through mucus.

In the intestinal barrier, one also find Peyer's patch. Peyer's patch are organized lymphoid nodules located in the ileum (follicle associated epithelium FAE), they contain specialized cells called M Cells which plays a role in the immune response. These cells deliver the antigen absorbed by the oral route to antigen presenting cells. There are two main reason to target M cells with nanocarriers via the oral route: either to stimulate a immune response and thus obtaining an oral vaccine directed versus antigen such as enterotoxins either to obtain drug absorption into lymph which is a mean to escape the liver first pass effect often responsible for a loss a bioavailability. If targeting M cells with functionalized nanocarriers is sought, one has to use a model featuring M cells to characterize these formulations. Regular CACO-2 models don't have M-cells. In the literature FAE models have been described and used to study the uptake of nanoparticles [56] designed to stimulate the immune response.

c. *In Vivo* or *Ex Vivo* Studies

In order to better mimic *in vivo* conditions, other more sophisticated models can be used. In these models the mucus is present. The everted sac model has been compared to CACO-2 cells to evaluate the absorption process of proliposomal beads containing cromolyn [57]. The transport pattern found with both models were not surprisingly different: after incubation over CACO-2 cells the transport was very effective and fast during the first 10 min and slowed down to reach a plateau after 200 min, whereas after incubation in the everted sacs the transport was first slow and accelerated with time. The cumulative amount absorbed is difficult to compare because the units are different.

In a model closer to *in vivo* condition, the single pass intestinal perfusion (SPIP), three formulations were compared by Liu *et al.* [28]: anionic solid lipid nanoparticles (SLNs), cationic SLNs and liposomes. The experiments were also performed on CACO-2 cells. In both models, SLNs displayed a better permeability than liposomes, but in SPIP anionic transport was the highest and in CACO-2 cationic transport was the highest.

In our research group, transport studies of paclitaxel lipid nanocapsules have been performed on CACO-2 cells [9] and were compared to the results obtained after a pharmacokinetic study, in the rat, involving the same drug delivery systems [27]. The permeability of paclitaxel across CACO-2 monolayers was enhanced by a threefold factor as this anticancer drug was encapsulated in lipid nanocapsules in comparison to its commercial formulation (Taxol®). In the pharmacokinetic study it was found that the bioavailability of paclitaxel after oral administration was tripled if encapsulated in lipid nanocapsules in comparison to Taxol®. Thus, for this example, a good correlation between CACO-2 permeability measurement and bioavailability observed in the rat was found. This particular result may be due to the fact that bioavailability of paclitaxel is impaired by its poor solubility and its efflux by the P-glycoprotein. The encapsulation in nanocapsules limited these drawbacks and thus the permeability enhancement was directly correlated with a better bioavailability. This good correlation between CACO-2 and *in vivo* experiment is not observed in the case of strong interaction with mucus or if there is an important hepatic first pass effect.

Conclusion

CACO-2 cells are a good screening tool to study permeability of drug encapsulated in nanocarriers. This model is fast and reliable if the set-up and validation process are rigorous and the culture conditions are standardized. It is possible to perform mechanistic studies in order to investigate the relative part of paracellular or transcellular pathways or the passive or active transports. It is even possible to study carrier-mediated endocytosis, by using specific inhibitors. It is also possible to study the effect of the efflux pump P-gP [21] or the presystemic metabolism mediated by cytochroms. But this model is not perfect, even if subclone such as TC7 are used. Nanocarriers have to diffuse through mucus to enter the enterocyte and the CACO-2 model lacks this feature. The absorption process can also occur via paracellular pathway and CACO-2 tight junction phenotype is closer to the colonic than the intestinal phenotype. Other structures are present in the intestinal barrier such as M-cells and are not present in CACO-2 cells. Co-culture of CACO-2 cells with other cell line has permitted to address some of these issues. In conclusion, CACO-2 cell model remain a necessary and central tool to investigate permeability of drug encapsulated in nanocarriers but it has to be associated with other complementary experiments to evaluate the interest of new drug delivery devices for oral drug administration.

ACKNOWLEDGMENT

Chapter reviewed by: Arnaud Beduneau, University of Franche-Comté, France & Anne Sapin, University of Nancy, France.

BIBLIOGRAPHY

[1] Sambuy, Y., et al., The CACO-2 cell line as a model of the intestinal barrier: influence of cell and culture-related factors on CACO-2 cell functional characteristics. *Cell Biol Toxicol,* 2005. 21(1): p. 1-26.

[2] Artursson, P. and R.T. Borchardt, Intestinal drug absorption and metabolism in cell cultures: CACO-2 and beyond. *Pharm Res,* 1997. 14(12): p. 1655-8.

[3] Artursson, P. and J. Karlsson, Correlation between oral drug absorption in humans and apparent drug permeability coefficients in human intestinal epithelial (CACO-2) cells. *Biochem Biophys Res Commun*, 1991. 175(3): p. 880-5.

[4] Yamashita, S., et al., Optimized conditions for prediction of intestinal drug permeability using CACO-2 cells. *Eur J Pharm Sci*, 2000. 10(3): p. 195-204.

[5] Konsoula, R. and F.A. Barile, Correlation of in vitro cytotoxicity with paracellular permeability in CACO-2 cells. *Toxicol In Vitro*, 2005. 19(5): p. 675-84.

[6] Delie, F., Evaluation of nano- and microparticle uptake by the gastrointestinal tract. *Advanced Drug Delivery Reviews*, 1998. 34(2-3): p. 221-233.

[7] Conner, S.D. and S.L. Schmid, Regulated portals of entry into the cell. *Nature*, 2003. 422(6927): p. 37-44.

[8] Snoeck, V., B. Goddeeris, and E. Cox, The role of enterocytes in the intestinal barrier function and antigen uptake. *Microbes and Infection,* 2005. 7(7-8): p. 997-1004.

[9] Roger, E., et al., Lipid nanocarriers improve paclitaxel transport throughout human intestinal epithelial cells by using vesicle-mediated transcytosis. *Journal of Controlled Release,* 2009. 140(2): p. 174-181.

[10] Ragnarsson, E.G.E., et al., Yersinia pseudotuberculosis induces transcytosis of nanoparticles across human intestinal villus epithelium via invasin-dependent macropinocytosis. *Laboratory Investigation,* 2008. 88(11): p. 1215-1226.

[11] Goldberg, D.S., H. Ghandehari, and P.W. Swaan, Cellular entry of G3.5 Poly (amido amine) dendrimers by clathrin- and dynamin-dependent endocytosis promotes tight junctional opening in intestinal epithelia. *Pharmaceutical Research*, 2010. 27(8): p. 1547-1557.

[12] Zha, L.Y., et al., Chromium nanoparticle exhibits higher absorption efficiency than chromium picolinate and chromium chloride in CACO-2 cell monolayers. *J Anim Physiol Anim Nutr* (Berl), 2008. 92(2): p. 131-40.

[13] Song, K.H., S.J. Chung, and C.K. Shim, Preparation and evaluation of proliposomes containing salmon calcitonin. *J Control Release*, 2002. 84(1-2): p. 27-37.

[14] Zhang, Y., et al., Cyclosporin A-loaded poly(ethylene glycol)-b-poly(d,l-lactic acid) micelles: preparation, in vitro and in vivo characterization and transport mechanism across the intestinal barrier. *Mol Pharm*, 2010. 7(4): p. 1169-82.

[15] Varma, M.V., et al., P-glycoprotein inhibitors and their screening: a perspective from bioavailability enhancement. *Pharmacol Res,* 2003. 48(4): p. 347-59.

[16] Goto, M., et al., Decreased expression of P-glycoprotein during differentiation in the human intestinal cell line CACO-2. *Biochem Pharmacol,* 2003. 66(1): p. 163-70.

[17] Hosoya, K.I., K.J. Kim, and V.H. Lee, Age-dependent expression of P-glycoprotein gp170 in CACO-2 cell monolayers. *Pharm Res,* 1996. 13(6): p. 885-90.

[18] Buckingham, L.E., et al., Reversal of multi-drug resistance in vitro by fatty acid-PEG-fatty acid diesters. *Int J Cancer,* 1996. 65(1): p. 74-9.

[19] Constantinides, P.P. and K.M. Wasan, Lipid formulation strategies for enhancing intestinal transport and absorption of P-glycoprotein (P-gp) substrate drugs: in vitro/in vivo case studies. *J Pharm Sci,* 2007. 96(2): p. 235-48.

[20] Coon, J.S., et al., Solutol HS 15, nontoxic polyoxyethylene esters of 12-hydroxystearic acid, reverses multidrug resistance. *Cancer Res,* 1991. 51(3): p. 897-902.

[21] Roger, E., et al., Reciprocal competition between lipid nanocapsules and P-gp for paclitaxel transport across CACO-2 cells. *Eur J Pharm Sci*, 2010. 40(5): p. 422-9.

[22] Risovic, V., et al., Potential mechanisms by which Peceol increases the gastrointestinal absorption of amphotericin B. *Drug Dev Ind Pharm*, 2004. 30(7): p. 767-74.

[23] Chen, F., et al., In vitro and in vivo study of N-trimethyl chitosan nanoparticles for oral protein delivery. *Int J Pharm*, 2008. 349(1-2): p. 226-33.

[24] Degim, Z., et al., The effect of various liposome formulations on insulin penetration across CACO-2 cell monolayer. *Life Sciences,* 2004. 75(23): p. 2819-2827.

[25] Wang, Y., et al., In vitro and in vivo evaluation of silybin nanosuspensions for oral and intravenous delivery. *Nanotechnology*, 2010. 21(15): p. 155104.

[26] Li, H., et al., Polyethylene glycol-coated liposomes for oral delivery of recombinant human epidermal growth factor. *International Journal of Pharmaceutics,* 2003. 258(1-2): p. 11-19.

[27] Peltier, S., et al., Enhanced oral paclitaxel bioavailability after administration of paclitaxel-loaded lipid nanocapsules. *Pharm Res*, 2006. 23(6): p. 1243-50.

[28] Liu, C., et al., In vitro and in vivo studies of lipid-based nanocarriers for oral N3-o-toluyl-fluorouracil delivery. *Drug Deliv,* 2010. 17(5): p. 352-63.

[29] Roger, E., et al., Biopharmaceutical parameters to consider in order to alter the fate of nanocarriers after oral delivery. *Nanomedicine*, 2010. 5(2): p. 287-306.

[30] Jevprasesphant, R., et al., Engineering of dendrimer surfaces to enhance transepithelial transport and reduce cytotoxicity. *Pharm Res*, 2003. 20(10): p. 1543-50.

[31] El-Sayed, M., et al., Transport mechanism(s) of poly (amidoamine) dendrimers across CACO-2 cell monolayers. Int J Pharm, 2003. 265(1-2): p. 151-7.

[32] Kitchens, K.M., et al., Endocytosis and interaction of poly (amidoamine) dendrimers with CACO-2 cells. *Pharmaceutical Research,* 2007. 24(11): p. 2138-2145.

[33] Kitchens, K.M., et al., Endocytosis inhibitors prevent poly(amidoamine) dendrimer internalization and permeability across CACO-2 cells. *Mol Pharm,* 2008. 5(2): p. 364-9.

[34] Sezgin, Z., N. Yuksel, and T. Baykara, Investigation of pluronic and PEG-PE micelles as carriers of meso-tetraphenyl porphine for oral administration. *Int J Pharm,* 2007. 332(1-2): p. 161-7.

[35] Lin, Y.H., et al., Preparation and characterization of nanoparticles shelled with chitosan for oral insulin delivery. *Biomacromolecules*, 2007. 8(1): p. 146-52.

[36] Lin, Y.-H., et al., Multi-ion-crosslinked nanoparticles with pH-responsive characteristics for oral delivery of protein drugs. *Journal of Controlled Release*, 2008. 132(2): p. 141-149.

[37] Ma, Z. and L.Y. Lim, Uptake of chitosan and associated insulin in CACO-2 cell monolayers: a comparison between chitosan molecules and chitosan nanoparticles. *Pharm Res,* 2003. 20(11): p. 1812-9.

[38] Lin, Y.H., et al., Novel nanoparticles for oral insulin delivery via the paracellular pathway. *Nanotechnology*, 2007. 18(10).

[39] Gao, P., et al., Development of a supersaturable SEDDS (S-SEDDS) formulation of paclitaxel with improved oral bioavailability. *J Pharm Sci,* 2003. 92(12): p. 2386-98.

[40] Zhang, Z. and S.S. Feng, Self-assembled nanoparticles of poly(lactide)--Vitamin E TPGS copolymers for oral chemotherapy. *Int J Pharm,* 2006. 324(2): p. 191-8.

[41] Batrakova, E.V., et al., Effects of pluronic P85 unimers and micelles on drug permeability in polarized BBMEC and CACO-2 cells. *Pharm Res*, 1998. 15(10): p. 1525-32.

[42] Jevprasesphant, R., et al., Transport of dendrimer nanocarriers through epithelial cells via the transcellular route. *J Control Release*, 2004. 97(2): p. 259-67.

[43] Win, K.Y. and S.S. Feng, Effects of particle size and surface coating on cellular uptake of polymeric nanoparticles for oral delivery of anticancer drugs. *Biomaterials*, 2005. 26(15): p. 2713-22.

[44] Mathot, F., et al., Transport mechanisms of mmePEG750P(CL-co-TMC) polymeric micelles across the intestinal barrier. *J Control Release,* 2007. 124(3): p. 134-43.

[45] Kitchens, K.M., et al., Transport of poly(amidoamine) dendrimers across CACO-2 cell monolayers: Influence of size, charge and fluorescent labeling. *Pharmaceutical Research,* 2006. 23(12): p. 2818-2826.

[46] Kitchens, K.M., M.E.H. El-Sayed, and H. Ghandehari, Transepithelial and endothelial transport of poly (amidoamine) dendrimers. *Advanced Drug Delivery Reviews,* 2005. 57(15): p. 2163-2176.

[47] Kansy, M., F. Senner, and K. Gubernator, Physicochemical high throughput screening: parallel artificial membrane permeation assay in the description of passive absorption processes. *J Med Chem,* 1998. 41(7): p. 1007-10.

[48] Sandhya, K., G. Devi, and Mathew, *Liposomal Formulations of Serratiopeptidase: In Vitro Studies Using PAMPA and CACO-2 Models. molecular pharmaceutics,* 2008. 5(1): p. 92-97.

[49] Hiremath, P.S., K.S. Soppimath, and G.V. Betageri, Proliposomes of exemestane for improved oral delivery: formulation and in vitro evaluation using PAMPA, CACO-2 and rat intestine. *Int J Pharm,* 2009. 380(1-2): p. 96-104.

[50] Turco, L., et al., CACO-2/TC7 cell line characterization for intestinal absorption: how reliable is this in vitro model for the prediction of the oral dose fraction absorbed in human? *Toxicol In Vitro,* 2011. 25(1): p. 13-20.

[51] Lai, S.K., Y.Y. Wang, and J. Hanes, Mucus-penetrating nanoparticles for drug and gene delivery to mucosal tissues. *Adv Drug Deliv Rev*, 2009. 61(2): p. 158-71.

[52] Behrens, I., et al., Transport of lipophilic drug molecules in a new mucus-secreting cell culture model based on HT29-MTX cells. *Pharm Res*, 2001. 18(8): p. 1138-45.

[53] Chen, X.M., I. Elisia, and D.D. Kitts, Defining conditions for the co-culture of CACO-2 and HT29-MTX cells using Taguchi design. *J Pharmacol Toxicol Methods,* 2010. 61(3): p. 334-42.

[54] Prego, C., et al., Efficacy and mechanism of action of chitosan nanocapsules for oral peptide delivery. *Pharm Res*, 2006. 23(3): p. 549-56.

[55] Dong, Y. and S.S. Feng, Poly(d,l-lactide-co-glycolide)/montmorillonite nanoparticles for oral delivery of anticancer drugs. *Biomaterials*, 2005. 26(30): p. 6068-76.

[56] des Rieux, A., et al., Transport of nanoparticles across an in vitro model of the human intestinal follicle associated epithelium. *Eur J Pharm Sci,* 2005. 25(4-5): p. 455-65.

[57] Deshmukh, D.D., W.R. Ravis, and G.V. Betageri, Improved delivery of cromolyn from oral proliposomal beads. *Int J Pharm,* 2008. 358(1-2): p. 128-36.

In: CACO-2 Cells and their Uses
Editor: Megan A. Schulz, pp. 181-196
ISBN 978-1-61209-917-0

Chapter 11

CACO-2 CELLS AS A MODEL TO STUDY THE INTESTINAL EFFECTS OF NON-ABSORBABLE PHYTOCHEMICALS

***A. G. Erlejman*[1], *C. G. Fraga*[2,3] *and P. I. Oteiza*[*3,4]**

[1]Departament of Biological Chemistry, School of Exacts and Natural Sciences, University of Buenos Aires-CONICET, Buenos Aires, Argentina.
[2]Physical Chemistry-PRALIB, School of Pharmacy and Biochemistry, University of Buenos Aires-CONICET, Buenos Aires, Argentina.
[3]Department of Nutrition,
[4]Department of Environmental Toxicology, University of California, Davis, CA, USA

ABSTRACT

Diets rich in fruits and vegetables are associated with a lower risk for colorectal cancer (CRC). Fruits and vegetables contain procyanidins, oligomers of flavan-3-ols that are among the most abundant polyphenols in human diets. Large procyanidins, i.e. composed of more than 3 monomer units (LPCA), are not transported inside cells. However, LPCA could act at the gastrointestinal level preventing chronic inflammation and the associated increased risk to develop CRC. CACO-2 cells is the model of intestinal epithelium that we have used to study the mechanisms through which LPCA could act protecting the integrity of the intestinal epithelial barrier, preventing oxidation and the triggering of inflammatory signals.We observed that LPCA bind to the plasmatic membrane and protect CACO-2 cells from oxidants, from the cytotoxic effects of deoxycholic acid (DCA), a secondary bile acid, and from inflammatory molecules, i.e. tumor necrosis alpha (TNFα). Mechanistically, it is plausible that LPCA act through: a) their free radical scavenging capacity; b) their direct interaction with membranes lipids, causing changes in membrane biophysics that would affect the activity of membrane-associated processes; and c) their interaction with membrane receptors that trigger intracellular responses. In summary, CACO-2 cells have proved to be an excellent model

* To whom correspondence should be addressed:Dr. Patricia I. Oteiza, Department of Nutrition, University of California, Davis, One Shields Av., Davis, CA, 95616, USA, Phone: 530-754-6074, Fax:530-752-8966, Email: poteiza@ucdavis.edu

to study the beneficial health effects of the non-absorbable LPCA. These findings further support the use of CACO-2 cells to identify the mechanisms of the intestinal actions of food components and/or pharmacological agents.

ABBREVIATIONS

AMVN, 2,2′-azobis(2,4-dimethylvaleronitrile); CRC, colorectal cancer; DCA, deoxycholicacid; DCDHF, 5-(and-6)-carboxy-20,70-dichlorodi hydro fluorescein diacetate; EC, (-)-epicatechin; Hex, Hexameric procyanidins;IBD, intestinal boweldiseases; iNOS, induciblenitricoxidesynthase; LDH, lacticde hydrogenase; LPCA, large procyanidins; TEER, transepithelial electrical resistance; TNFα, tumornecrosis factor alpha.

1. INTRODUCTION

Many large molecules present in foods are not absorbed and can accumulate in the intestine where they may: a) follow the transit and be excreted; b) be metabolized to smaller and absorbable compounds by the gut microflora; or c) interact with intestinal cells and generate a local or systemic response. Procyanidins, especially those composed of more than three units of flavan-3-ols (hereon defined as large procyanidins, LPCA) are examples of those large molecules. The use of CACO-2 cell monolayers allowed us to demonstrate that although LPCA are not transported through the intestinal ephitelium, they can still exert specific biological responses[1]. LPCA have previously shown a high affinity in their interaction with membrane lipids [2]. Thus, we hypothesized that through those interactions, LPCA could protect intestinal cells from different aggressors including secondary bile acids, bacterial toxins, inflammatory agents, and putative oxidants [3]. In part this protection would be mediated by the capacity of LPCAto regulate cell signaling initiated at the cell membrane level [4]. The above hypothesis was tested using differentiated CACO-2 cells as a model of intestinal epithelium.This chapter will summarize our findings supporting the concept that even when LPCAare not absorbed by intestinal cells they could provide health benefitsat the gastrointestinal tract.

2. PROCYANIDINS:STRUCTURE AND METABOLISM

Procyanidins are present in significant amounts in the human diet. Theirpresence is particularly high in cocoa, grapes, cranberries, apples, and their derived foods and beverages [5-6]. Procyanidins are oligomers of flavan-3-ol subunits, i.e. (+)-catechin and (-)-epicatechin (EC). Our studies have used procyanidins present in cocoa, which are primarily composed of units of EC linked by 4β→8 bonds (Figure 1). Procyanidins oligomerization size ranges between 2 and, about 10 subunits, and their chemical characteristics vary depending on the particular plant source.

Current evidence indicates that monomers and dimeric procyanidins are absorbed, while larger procyanidins have not been detected in plasma after their dietary consumption [7].

Consistently, procyanidins composed ofmore than 3 units are not transported in vitro through CACO-2 cell monolayers [1]. In humans, only procyanidin dimers seem to be absorbed intact, but at with much lower efficiency than the monomers [8]. In vitro, human fecal microflora metabolize dimericprocyanidin B2 ((-)-epicatechin-4beta-->8-(-)-epicatechin) into 24 compounds[9].

After the ingestion of a grape seed extract, the B2 dimer is absorbed and appears in blood [10], but is also found in measurable concentrations in rat colon [11]. *In vitro*, human colonic microflora metabolize procyanidins into phenylacetic, phenylpropionic and phenylvaleric acids suggesting that, at least in part, LPCA could be converted in the human gut to phenolic acids that could be absorbed as such [12]. However, the finding thatafter the consumption of LPCA-rich foods, the presence of phenolic acids is very low in rat urine, suggest that LPCA metabolization by the gut microflorawould not be extensive in vivo. Thus, non metabolized LPCA would not be absorbed and be present in the intestine where they could interact with the epithelial cells.

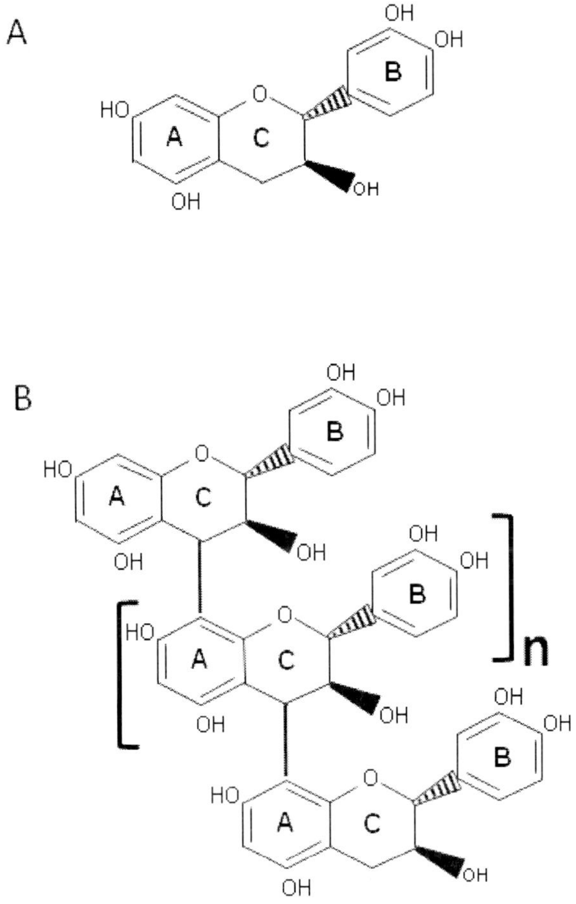

Figure 1. Chemical structures of A- (-)-epicatechin; and B-procyanidins built from epicatechin subunits linked by 4β→8 bonds. n: number of (-)-epicatechin units.

3. Procyanidins and the Health of the Gastrointestinal Tract

Inflammatory bowel diseases (IBD), including the severe and frequent Crohn's disease and ulcerative colitis, affect a large number of individuals in the United States and worldwide[13-14]. IBD are characterized by a chronic inflammation of the gastrointestinal tract[15] that is associated with an increased risk to develop colorectal cancer (CRC). CRC is the third cancer type causing death in the U.S.A. and worldwide. In IBD and CRC, both genetic and environmental components contribute to the development of the disease.

Chronic inflammation in IBDinvolves the overactivation of the immune system that promotes a cross-talk between immune and epithelial cells. Such interaction leads to a loss of the barrier integrity, which can allow the entrance of potentially damaging xenobiotics and antigens that initiate another cycle of immune cell activation [16].

Several transcription factors participate in the pathogenesis of IBD mediating those events, in a kind of "cyclic" propagation.The activation of NF-κB, NAFTc2, and STATs [17]in native immune cells at the lamina propialeads to the synthesis of cytokines and chemokines, that attract and activate other immune cells (granulocyte, macrophages). These immune cells also lead to an increase in the production and tissue steady state levels of oxidants. Both pro-inflammatory mediators and oxidants canpromote alterations in the intestinal barrier integrity, further activate redox sensitive transcription factors (NF-κB, NFAT, STATs)in epithelial and immune cells,completing the cycle of inflammation and intestinal damage in IBD. Significantly, both ulcerative cholitis and CRC present similar alterations in a number of signaling pathways (STAT3 and NF-κB, β-catenin, Wnt, p53, TGFβ) [18].

A population study showed that the dietary intake of procyanidins is inversely related to the risk of CRC [19]. While the ingestion of monomers did not affect the CRC risk, the ingestion of procyanidins with either two or more EC units was associated to protective effects [19]. In addition, procyanidins regulate signals involved in IBD and carcinogenesis[20]. In this regard, we described that dimeric and hexameric procyanidins isolated from cocoa inhibit NF-κB in CACO-2 cells and lymphocytes under proinflammatory conditions [4, 21-23]. Dimericprocyanidin B1 and B2, but not dimers A1 and A2, inhibit NF-κB-dependent IL-2 production in Jurkat T cells [22-23]. B2 also inhibits constitutively active NF-κB in Hodgkin's lymphoma cells, decreasing cell viability, and the production of IL-6, TNFα, and Rantes[21]. As described in other sections of this chapter, an hexameric LPCA fraction protected CACO-2 cell monolayers from bile- and hydrophobic oxidant-induced oxidative stress, and barrier permeabilization[3]. Hexameric procyanidins (Hex) also inhibited TNFα-induced NF-κB activation and the associated inducible nitric oxide synthase (iNOS) expression [4]. In animal models of chronic inflammation, plant extracts with high procyanidin content (apple and grape seed extracts) have anti-inflammatory, and protective actions in models of ulcerative colitis and CRC [20, 24-25].

4. Interactions of Procyanidins with CACO-2 Cell Membranes

We initially investigated, using synthetic lipid bilayers (liposomes), the hypothesis that LPCA could have biological actions through their adsorption to the cell membrane. LPCA could adsorb to membranes through their interactions with the lipid polar head groups. These interactions could confer antioxidant and cytoprotective properties to LPCA by preventing the access, or interacting with, oxidants and membrane-disrupting agents before they could reach the bilayer. Procyanidins of 2 to 6 EC units protected liposomes (composed of phosphatidylcholine and phosphatidylserine) from oxidation induced by a lipid soluble pro-oxidant [2]. Their antioxidant action positively correlated with the number of monomeric units in the procyanidin. LPCA also protected membranes from the transition of lipid bilayer to micelles induced by a non-ionic detergent [2, 26]. These protective effects cannot be ascribed to LPCA-induced changes in membrane biophysics. In this regard, LPCA did not induce lateral phase separation and changes in membrane fluidity (evaluated with the probe 1,6-diphenyl-1,3,5-hexatriene)[2]. In phosphatidylcholine liposomes, dimeric procyanidins, and at a much larger extent, hexameric procyanidins, increase the fluorescence polarization of a probe sensing areas close to the surface region of the bilayer [26]. Hexameric LPCA did not affect the fluorescence polarization of a probe localized in the hydrophobic core of the membrane. The above evidence suggests that LPCA could adsorb to the polar region of biological membranes probably through hydrogen bonds. We subsequently investigated the capacity of a hexameric LCPA to interact with polarized CACO-2 cell monolayers, used as the model to mimic in vitro an intestinal epithelium. We measured the capacity of Hex to modify the transepithelial electrical resistance (TEER) as a parameter of LPCA-membrane interactions. Hex caused an increase in TEER between 30 and 120 min of its addition to the incubation medium (Figure 2) [3]. This increase was reversed to initial values within 60 min when Hex was removed from the medium after 30 min of pre-incubation. Those findings indicate that LPCA is not causing an actual change in the cell monolayer permeability, but that affects TEER by adsorbing to the cell membrane. Importantly, they also show that the binding of LPCA to CACO-2 cells membranes is reversible.

Besides the finding that LPCA adsorption does not cause adverse effects to the barrier integrity, we also observed that it does not affect normal cellular and paracellular transport [3]. Similarly, it was previously shown in CACO-2 cells, that only monomers, dimers, and trimers are transported through the cell monolayer [1]. No components of an extract of mixed LPCA (average polymerization degree of 6) were transported, but they caused an increase in TEER [1]. Thus, both in synthetic membranes and CACO-2 cell membranes, LPCA can reversibly adsorb to the surface of the bilayer without affecting membrane biophysics and normal metabolite transport.

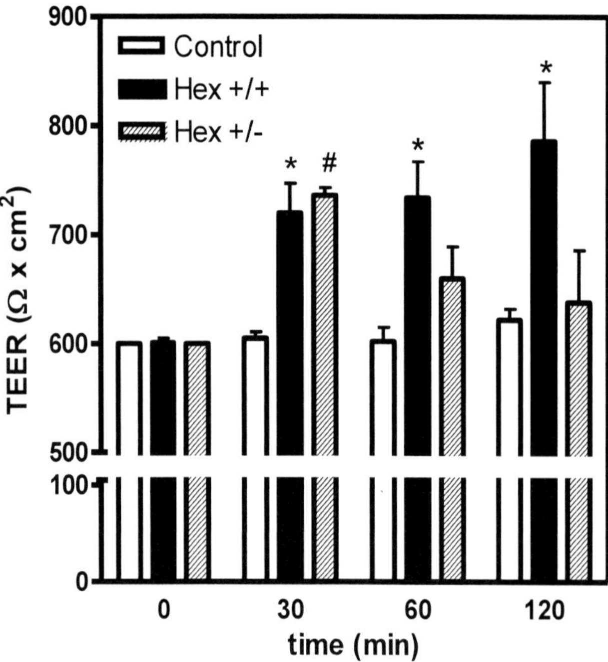

Figure 2. Interactions of hexameric LPCA with CACO-2 cell membranes. Measurement of the transepithelial electrical resistance (TEER) was used to evaluate the interactions of hexameric LPCA (Hex) with CACO-2 cells membranes. CACO-2 cells were incubated at 37°C for 120 min in media without (control)(empty bars); or with the addition of 20 μMhexameric LPCA (Hex +/+) (full bars); or incubated for 30 min in media added with 20μM Hex and for further 30 or 90 min in Hex free-media (Hex +/-) (dashed bars). Results are shown as mean ±SE of at least three independent experiments. *,#Significantly different from the control group at the corresponding time point ($p<0.05$). Adapted from Erlejman et al. [3].

5. Procyanidins ProtectCACO-2 Cellsfrom Oxidation and Permeabilization

The intestinal epithelium represents the largest interface between the host and the environment. Moreover it is the main organ responsible of nutrient absorption, in the regulation of electrolytic interchange, and preventing the entrance of potential antigens. All these functions depend on the integrity of the intestinal epithelial barrier. The conservation of intercellular tight junctions contributes to maintaining the barrier homeostasis. Depending on the magnitude and type of the antigen challenge, the intestinal immune response could be tolerogenic, or it could initiate and/or perpetuate chronic inflammation. Chronic inflammation can alter the paracellular and transcellular permeability of the barrier [16, 27], which constitute one of the most critical events leading to intestinal dysfunction. Aspreviously mentioned, numerousdiseases such as IBD, CRC, celiac disease, alcoholic liver disease, fatty acid liver disease, food allergies, and acute pancreatitis are associated with intestinal epithelial barrier defects [18, 28]. Stressing its relevance to IBD, it has been proposed the evaluation of the intestinal permeability as a clinical endpoint in the development of futureIBD therapies[29]. The factors that have been associated with alterations of the intestinal barrier include, among others,nonsteroidal anti-inflammatory drugs, cytokines

(TNFα, IL-1β) [30-31], and oxidative stress. The proposed molecular mechanisms include adenosine triphosphate depletion, activation of select transcription factors that modulate genes involved in tight junction regulation [30-31], and damage to cytoskeletal proteins that regulate tight junctions [28].

Secondary bile acids are formed after enzymatic deconjugation and dehydroxylation of primary bile acids in the large bowel by anaerobic bacteria. They can reach high concentrations in the enterohepatic circulation and in the intestine[32]. It has been proposed that these compounds, in particular deoxycholic acid (DCA) can promote intestinal tumorogenesis[33-34]. In fact, epidemiological studies have associated a higher colonic cancer risk with the fecal bile acid concentration[35-36]. Mechanistically, DCA perturbs the cell membrane affecting its lipid distribution and physical properties, and promotes an increase in cell oxidant production[37-38]. Furthermore, through alterations of membrane microdomains, DCA can trigger different cell signaling cascades including protein kinase C [39-40], the EGFR/Raf-1/ERK[41-42]pathway, and transcription factors NF-κB [43-45], and AP-1 [44, 46].

We evaluated the capacity of anhexameric LCPA fraction to prevent the adverse effects of DCA on CACO-2 cell monolayers. DCA was highly cytotoxic to CACO-2 cells, generating oxidants and altering the permeability of polarized CACO-2 cells monolayers[3].As an evidence of DCA cytotoxicity, it caused a time- and dose-dependent increase in the release of the cytosolic enzyme lactic dehydrogenase (LDH) and a decrease in cell viability. The hexameric LPCA fraction decreased DCA-induced LDH release in a dose-depend manner, being the protective effect significant in all the range of concentrations tested (2.5-20 μM) [3]. LPCA also prevented the decrease in cell viability within 6 h of incubation with DCA[3].

It has been previously shown that DCA in part causes intestinal epithelium dysfunction by increasing oxidant production [47]. Moreover, other bile salts affect CACO-2 monolayer permeability by mechanisms involving oxidants[48]. We observed that 0.2 mM DCA causes an increase in the steady state levels of oxidants[3]. Cell oxidant levels were evaluated measuring the oxidation of the probe 5-(and-6)-carboxy-2′7′-dichlorodihydrofluorescein diacetate (DCDHF). This probe crosses the cell membrane, and fluoresces when it is oxidized. After 1h of treatment with DCA (0.2mM) cell oxidants were 50% higher than in untreated cells. Co-treatment with hexameric LPCA caused a dose-dependent reduction in cell oxidant levels[3]. The same protective effect was observed after removal of the LPCA from the cell culture media (Figure 3.A).

We next investigated if the hexameric LPCA could act preventing cell oxidant production generated by a lipophilic oxidant, 2,2′-azobis(2,4-dimethylvaleronitrile) (AMVN). Following 60 min of incubation with 10 mM AMVN, cell oxidant levels raised by 100% over basal levels. Also in this case, the hexameric LPCA acted inhibiting in a dose–depend manner the increase of oxidants, even after LPCA removal from the cell culture media (Figure 3.B)[3].

The above results demonstrate the capacity of the non absorbable LPCA to protect intestinal cells from secondary bile acid-mediated cytotoxicity and from oxidative stress. These effects were observed even after removal of the LPCA from the incubation media, which coincides with a gradual TEER decrease after LPCA removal (Figure 2). Thisfinding stresses that LPCA binding to cell membranes, although reversible, can provide protection beyond the time of direct contact of these compounds with the cell membrane.

DCA causes a redistribution of the tight junction protein ZO-1 from cell-cell contacts into the cytoplasm causing epithelial barrier permeabilization[3]. ZO proteins are peripherally associated membrane proteins which name derives from the initials of zonulaoccludens (for a review on tight junction proteins, see [49]).

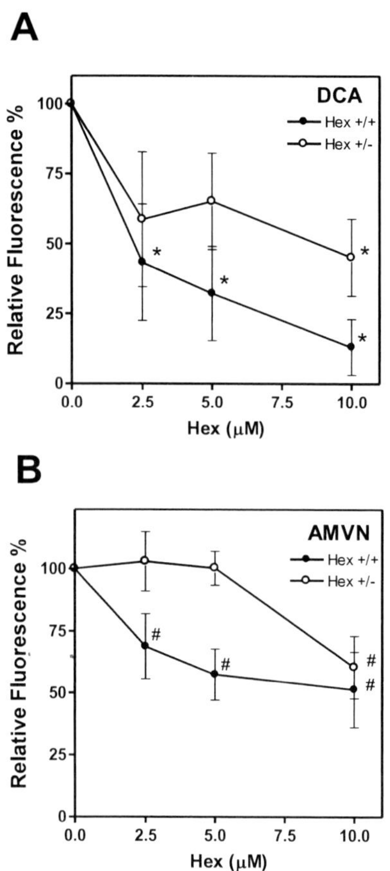

Figure 3. Effects ofhexameric LPCA on CACO-2 cell oxidant levels. Oxidant levels were evaluated by the fluorescence of DCDHF incorporated into the CACO-2 cells. Cells were preincubated in media without hexameric LPCA (Hex), or preincubated for 30 min in the presence of 2.5-10 μM Hex, and then incubated for 60 min with: A) 0.2 mM DCA; or B) 10 mM AMVN in (Hex +/+ or Hex +/- conditions, see Fig. 2 legend). Fluorescence of cells incubated in the absence of additions was considered as 100%. Results are shown as mean ±SE of four independent experiments. *Significantly different compared to the value for the cells incubated without Hex ($p<0.01$);[#]significantly different compared to the value for the cells incubated without Hex ($p<0.05$). Adapted from Erlejman et al. [3].

These proteins interact together and anchor membrane proteins like claudins, occludin and JAMs to the actin cytoskeleton. The distribution of ZO-1 in CACO-2 cells monolayer evaluated by fluorescence microscopy showed the typical location at the cell boundaries[3]. After the treatment of cells with 0.2 mM DCA for 6 h there was adecrease in the amount of ZO-1 in areas of cell-cell contact and an increase in cytosolic ZO-1[3]. These changes in ZO-1 localization can be associated with alterations in paracellular permeability. The permeability through the paracellular pathway can be evaluated by two methods, the measurement of TEER, and of the flux of tracers that could be easily detected (i.e fluorescent probes) [49]. As a measure of paracellular transport were tested the transport of fluorescein sulfonic acid (FSA). An increase in FSA transport reflects a loss of the selective permeability caused by perturbations in CACO-2 monolayer integrity.Both, DCA and AMVN altered the barrier permeability causing an increase in FSA paracellular transport. Co-treatment with hexameric LPCA partially prevents the increase in FSA clearance induced either by DCA or AMVN (Table 1).

Table 1. Effect of hexameric LPCA on DCA-induced alterations on CACO-2 barrier permeability

	Clearance FSA (nl/h/cm^2)	
	-Hex	+Hex
DCA	1.44 ± 0.08	1.24 ± 0.06 *
AMVN	2.09 ± 0.08	1.41 ± 0.12 #

Table1. Cells were preincubated in theabsence (-Hex) or presence of 10 μM Hex (+Hex) for 30 min and then added with 0.2 mMDCA or 10 mMAMVN for 6 and 1 h, respectively. FSAclearance was measuring as the apical-to basolateral clearance of FSA as previously described [3]. Results are expressed as time of increment considering the basal level (without treatment) equal to one. Result are shown as mean ±SE of four independent experiments. *Significantly different compared to the value for the cells incubated without Hex ($p<0.04$). #Significantly different compared to the value for the cells incubated without Hex ($p<0.007$). This table was adapted from Erlejman et al. [3].

These results demonstrate that hexameric LPCA can prevent DCA-mediated increase in cell oxidants, and oxidant-mediated alterations in the permeability of CACO-2 monolayers. The above findings further support the concept that, at the intestinal level, LPCA could participate in the protection of the gastrointestinal tract from the damaging effects of oxidants and secondary bile acids.

6. Procyanidins ProtectCACO-2 Cells from Inflammation

High levels of TNFα are observed in IBD patients, and anti-TNFα antibodies are used in the treatment of IBD [50-51]. These facts stressthe central role that TNFα plays in the pathogenesis of IBD. TNFα triggers the activation of transcription factor NF-κB that, although physiologically involved in the immune response, also contributes to chronic inflammation and intestinal barrier permeabilization. One consequence of TNFα-induced NF-κB activation is the upregulation of myosin light chain kinase, which triggers an increase in tight junction permeability [52-53]. Also dependent on NF-κB, two components of the organizer of tight junctions and polarity, the polarity complex (atypical protein kinase C [aPKC]-PAR6-PAR3), are down regulated by TNFα[54]. These effects were observed both in intestinal epithelial cells, and also in vivo during intestinal inflammation.

Barrier dysfunction, activation and migration of immune cells at the lamina propia, the release of pro-inflammatory mediators and the activation of epithelial cells create a cycle of chronic inflammation and tissue damage in which NF-κBhas a critical role[55-

56].Furthermore, the presence of inflammatory mediators, such as TNFα, are well described in the microenvironment of neoplasic transformation, playing a central role in the progression of inflammation-related tumors [57].

Given the relevance of TNFα in IBD and CRC, we studied the potential anti-inflammatory actions of LPCA characterizing the capacity of LPCA to inhibit TNFα-induced NF-κB activation.We used as a model of intestinal inflammation CACO-2 cells exposed to high TNFα concentrations. The activation of NF-κB by TNFα involves the canonical activation pathway. NF-κBresides in the cytosol as homo and heterodimers of the Rel/NF-κB family of proteins. Binding to the inhibitory IκB proteins masks the NF-κB nuclear localization sequence keeping the transcription factor in its cytosolic localization[58].

The activation of specific IκB kinases (IKK), leads to the phosphorylation of IkB, which targets the protein for ubiquitination and subsequent degradation by the proteasome [59].The activeNF-κBsubsequently translocates to the nucleus,andbinds to the promoter of several genes involved, among others, in inflammation and barrier permeabilization.

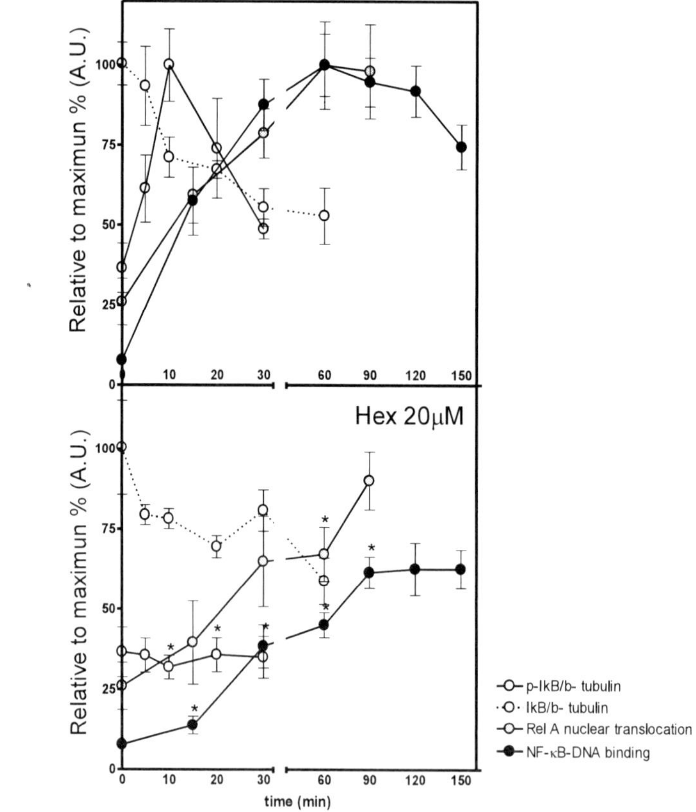

Figure 4. Effects of hexameric LPCA on inflammation: inhibition of TNFα-induced NF-κB activation. Cells were incubated in the absence (upper panel), or presence (bottom panel) of 20 μM Hex, and then incubated with 10 ng/ml TNFα for up to 150 min. The levels of p-IκB, IκB and Rel A nuclear translocation were evaluated by Western Blot. NF-κB-DNA binding was evaluated by Electrophoretic Mobility Shift Assay. Results are shown as mean ±SE of for least three independent experiments. *Significantly different compared to the value for the cells incubated without Hex at the corresponding time(upper panel) ($p<0.01$). Adapted from Erlejman et al.[4].

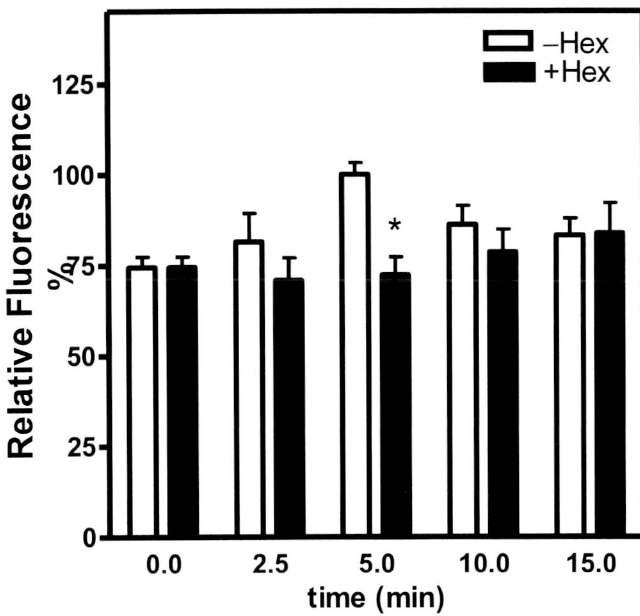

Figure 5. Effects of hexameric LPCA on TNFα-induced CACO-2 cells oxidation. Oxidant levels were evaluated by the fluorescence of DCDHF incorporated into the CACO-2 cells . Cells were preincubated in the absence (Control) (empty bars), or presence (Hex) (full bars) of 20 μM Hex, and then incubated with 10 ng/ml TNFα for further 2.5 to 15 min. Results are shown as mean ±SE of four independent experiments. *Significantly different compared to the value for the cells incubated without Hex ($p<0.005$). Adapted from Erlejman et al.[4].

In CACO-2 cells, TNFα (10 ng/ml) triggers the phosphorylation and degradation of IκBα that reaches a maximum after 10 and 20 min of incubation, respectively. (Figure 4 upper panel). Cells preincubated with hexameric LPCA and then co-treated with TNFα showed a significant inhibition of IκBphosporylation, and a delay in IκB degradation (Figure 4 bottom panel). The following step in the activation cascade, NF-κB nuclear translocation, was followed as the amount of the NF-κB protein Rel A present innuclear fractions by Western blot (Fig.4 upper panel). LPCA prevented the nuclear accumulation of Rel A induced by TNFα (Figure 4. bottom panel).NF-κB-DNA binding was evaluated by electrophoretic mobility shift assay(EMSA). In CACO-2 cells, TNFα-activated NF-κB was composed mainly by p50 and Rel A peptides [4]. The kinetic (15-120 min) of NF-κB-DNA nuclear binding showed that Hex also decreased TNFα-induced increase in NF-κB-DNA binding (Figure 4).

Importantly, the hexameric LPCA inhibited TNFα-induced increase in the expression of the NF-κB-inducible nitric oxide synthase (iNOS). The preincubation with Hex abolished the raise in iNOS expression [4]. The inhibitory actions of LPCA were not associated with their antioxidant activity. TNFα causes a transient increase of cellular oxidants, which is prevented by LPCA (Figure 5). However the antioxidants N-acetyl cysteine or α-lipolic acid (0.5 mM) although inhibited TNFα-induced oxidant increase, do not affect the increase in IκB phophorylation[4].

Very importantly, the inhibitory action of hexameric LPCA on TNFα- induced NF-κB activation was observed even after the removal of the LPCA from the media. This demonstrates once again that LPCA can transiently adsorb to CACO-2 cell membranes andinhibit TNFα-induced NF-κB activation through local membrane effects.

Conclusion

This chapter discussed the usefulness of differentiated CACO-2 cells as a model of normal intestinal epithelium to study the potential biological effects of phytochemicals that, although not absorbed, could have local effects at the gastrointestinal tract.In CACO-2 cells, we have studied the antioxidant, cytoprotective, and anti-inflammatory actions of LPCA. LPCA are phytochemicals present in large amounts in the human diet.Allthough their metabolism in the gut is not fully described; current evidence indicates that their degradation by the gut microflora would be limited, and that they could reach the colon in significant concentrations.

In synthetic membranes, LPCA adsorb to the lipid bilayer and have antioxidant and membrane protective effects. In CACO-2 cells, LPCA reversibly adsorb to the cell membrane and have cytoprotective and antioxidant action, preventing the loss of the intestinal epithelial barrier selective permeability. LPCA also protect CACO-2 cell monolayers from the adverse effects of inflammatory cytokines.

Thus, diets rich in LPCA could prevent and ameliorate chronic inflammatory processes at the gastrointestinal tract, and contribute to prevent the onset and progression of IBD and CRC.

Acknowledgments

This work was supported by grants from the University of California, Davis; and the CHNR-State of California Vitamin Price Fixing Consumer Settlement Fund, USA; and the University of Buenos Aires, UBACyT (2010-2013).AGE is Assistant Investigator, CGF is Principal Investigator, and PO is Corresponding Investigator of CONICET, Argentina.

References

[1] Deprez S, Mila I, Huneau JF, Tome D, Scalbert A. Transport of proanthocyanidin dimer, trimer, and polymer across monolayers of human intestinal epithelial CACO-2 cells. *Antioxid Redox Signal.* 2001 Dec;3(6):957-67.

[2] Verstraeten SV, Keen CL, Schmitz HH, Fraga CG, Oteiza PI. Flavan-3-ols and procyanidins protect liposomes against lipid oxidation and disruption of the bilayer structure. *Free Radic Biol Med.* 2003 Jan 1;34(1):84-92.

[3] Erlejman AG, Fraga CG, Oteiza PI. Procyanidins protect CACO-2 cells from bile acid- and oxidant-induced damage. *Free Radic Biol Med.* 2006 Oct 15;41(8):1247-56.

[4] Erlejman AG, Jaggers G, Fraga CG, Oteiza PI. TNFalpha-induced NF-kappaB activation and cell oxidant production are modulated by hexameric procyanidins in CACO-2 cells. *Arch Biochem Biophys.* 2008 Feb 5.

[5] Gu L, Kelm MA, Hammerstone JF, Beecher G, Holden J, Haytowitz D, et al. Concentrations of proanthocyanidins in common foods and estimations of normal consumption. *J Nutr*. 2004 Mar;134(3):613-7.

[6] Hammerstone JF, Lazarus SA, Schmitz HH. Procyanidin content and variation in some commonly consumed foods. *J Nutr*. 2000 Aug;130(8S Suppl):2086S-92S.

[7] Sano A, Yamakoshi J, Tokutake S, Tobe K, Kubota Y, Kikuchi M. Procyanidin B1 is detected in human serum after intake of proanthocyanidin-rich grape seed extract. *Biosci Biotechnol Biochem*. 2003 May;67(5):1140-3.

[8] Manach C, Williamson G, Morand C, Scalbert A, Remesy C. Bioavailability and bioefficacy of polyphenols in humans. I. Review of 97 bioavailability studies. *Am J Clin Nutr*. 2005 Jan;81(1 Suppl):230S-42S.

[9] Stoupi S, Williamson G, Drynan JW, Barron D, Clifford MN. Procyanidin B2 catabolism by human fecal microflora: partial characterization of 'dimeric' intermediates. *Arch Biochem Biophys*. 2010 Sep 1;501(1):73-8.

[10] Stoupi S, Williamson G, Viton F, Barron D, King LJ, Brown JE, et al. In vivo bioavailability, absorption, excretion, and pharmacokinetics of [14C]procyanidin B2 in male rats. *Drug Metab Dispos*. 2010 Feb;38(2):287-91.

[11] Tsang C, Auger C, Mullen W, Bornet A, Rouanet JM, Crozier A, et al. The absorption, metabolism and excretion of flavan-3-ols and procyanidins following the ingestion of a grape seed extract by rats. *Br J Nutr*. 2005 Aug;94(2):170-81.

[12] Deprez S, Brezillon C, Rabot S, Philippe C, Mila I, Lapierre C, et al. Polymeric proanthocyanidins are catabolized by human colonic microflora into low-molecular-weight phenolic acids. *J Nutr*. 2000 Nov;130(11):2733-8.

[13] Terry P, Giovannucci E, Michels KB, Bergkvist L, Hansen H, Holmberg L, et al. Fruit, vegetables, dietary fiber, and risk of colorectal cancer. *J Natl Cancer Inst*. 2001 Apr 4;93(7):525-33.

[14] Payne CM, Crowley-Skillicorn C, Bernstein C, Holubec H, Moyer MP, Bernstein H. Hydrophobic bile acid-induced micronuclei formation, mitotic perturbations, and decreases in spindle checkpoint proteins: Relevance to genomic instability in colon carcinogenesis. *Nutrition and Cancer*. 2010;62(6):825-40.

[15] Bouma G, Strober W. The immunological and genetic basis of inflammatory bowel disease. *Nat Rev Immunol*. 2003 Jul;3(7):521-33.

[16] Xavier RJ, Podolsky DK. Unravelling the pathogenesis of inflammatory bowel disease. *Nature*. 2007 Jul 26;448(7152):427-34.

[17] Mudter J, Neurath MF. Il-6 signaling in inflammatory bowel disease: pathophysiological role and clinical relevance. *Inflamm Bowel Dis*. 2007 Aug;13(8):1016-23.

[18] Terzic J, Grivennikov S, Karin E, Karin M. Inflammation and colon cancer. *Gastroenterology*. 2010 Jun;138(6):2101-14 e5.

[19] Rossi M, Negri E, Parpinel M, Lagiou P, Bosetti C, Talamini R, et al. Proanthocyanidins and the risk of colorectal cancer in Italy. *Cancer Causes Control*. 2010 Feb;21(2):243-50.

[20] Nandakumar V, Singh T, Katiyar SK. Multi-targeted prevention and therapy of cancer by proanthocyanidins. *Cancer Lett*. 2008 Oct 8;269(2):378-87.

[21] Mackenzie GG, Adamo AM, Decker NP, Oteiza PI. Dimeric procyanidin B2 inhibits constitutively active NF-kappaB in Hodgkin's lymphoma cells independently of the presence of IkappaB mutations. *Biochem Pharmacol.* 2008 Apr 1;75(7):1461-71.

[22] Mackenzie GG, Carrasquedo F, Delfino JM, Keen CL, Fraga CG, Oteiza PI. Epicatechin, catechin, and dimeric procyanidins inhibit PMA-induced NF-kappaB activation at multiple steps in Jurkat T cells. *Faseb J.* 2004 Feb;18(1):167-9.

[23] Mackenzie GG, Delfino JM, Keen CL, Fraga CG, Oteiza PI. Dimeric procyanidins are inhibitors of NF-kappaB-DNA binding. *Biochem Pharmacol.* 2009 Nov 1;78(9):1252-62.

[24] Gosse F, Roussi S, Guyot S, Schoenfelder A, Mann A, Bergerat JP, et al. Potentiation of apple procyanidin-triggered apoptosis by the polyamine oxidase inactivator MDL 72527 in human colon cancer-derived metastatic cells. *Int J Oncol.* 2006 Aug;29(2):423-8.

[25] Velmurugan B, Singh RP, Kaul N, Agarwal R, Agarwal C. Dietary feeding of grape seed extract prevents intestinal tumorigenesis in APCmin/+ mice. *Neoplasia.*Jan;12(1):95-102.

[26] Erlejman AG, Verstraeten SV, Fraga CG, Oteiza PI. The interaction of flavonoids with membranes: potential determinant of flavonoid antioxidant effects. *Free Radic Res.* 2004 Dec;38(12):1311-20.

[27] Menard S, Cerf-Bensussan N, Heyman M. Multiple facets of intestinal permeability and epithelial handling of dietary antigens. *Mucosal Immunol.* 2010 May;3(3):247-59.

[28] DeMeo MT, Mutlu EA, Keshavarzian A, Tobin MC. Intestinal permeation and gastrointestinal disease. *J Clin Gastroenterol.* 2002 Apr;34(4):385-96.

[29] Han X. Intestinal permeability as a clinical surrogate endpoint in the development of future Crohn's disease therapies. *Recent Pat Inflamm Allergy Drug Discov.* 2010;4(2):159-76.

[30] Ye D, Ma I, Ma TY. Molecular mechanism of tumor necrosis factor-alpha modulation of intestinal epithelial tight junction barrier. *Am J Physiol Gastrointest Liver Physiol.* 2006 Mar;290(3):G496-504.

[31] Al-Sadi R, Ye D, Said HM, Ma TY. IL-1beta-induced increase in intestinal epithelial tight junction permeability is mediated by MEKK-1 activation of canonical NF-kappaB pathway. *Am J Pathol.* 2010 Nov;177(5):2310-22.

[32] Ridlon JM, Kang DJ, Hylemon PB. Bile salt biotransformations by human intestinal bacteria. *J Lipid Res.* 2006 Feb;47(2):241-59.

[33] Mahmoud NN, Dannenberg AJ, Bilinski RT, Mestre JR, Chadburn A, Churchill M, et al. Administration of an unconjugated bile acid increases duodenal tumors in a murine model of familial adenomatous polyposis. *Carcinogenesis.* 1999 Feb;20(2):299-303.

[34] Nagengast FM, Grubben MJ, van Munster IP. Role of bile acids in colorectal carcinogenesis. *Eur J Cancer.* 1995 Jul-Aug;31A(7-8):1067-70.

[35] Imray CH, Radley S, Davis A, Barker G, Hendrickse CW, Donovan IA, et al. Faecal unconjugated bile acids in patients with colorectal cancer or polyps. *Gut.* 1992 Sep;33(9):1239-45.

[36] Kishida T, Taguchi F, Feng L, Tatsuguchi A, Sato J, Fujimori S, et al. Analysis of bile acids in colon residual liquid or fecal material in patients with colorectal neoplasia and control subjects. *J Gastroenterol.* 1997 Jun;32(3):306-11.

[37] Jean-Louis S, Akare S, Ali MA, Mash EA, Jr., Meuillet E, Martinez JD. Deoxycholic acid induces intracellular signaling through membrane perturbations. *J Biol Chem*. 2006 May 26;281(21):14948-60.

[38] Akare S, Martinez JD. Bile acid induces hydrophobicity-dependent membrane alterations. *Biochim Biophys Acta*. 2005 Jun 15;1735(1):59-67.

[39] Redlak MJ, Power JJ, Miller TA. Protein kinase C involvement in deoxycholate-induced apoptosis in human gastric cells. *Dig Dis Sci*. 2006 May;51(5):834-43.

[40] Lau BW, Colella M, Ruder WC, Ranieri M, Curci S, Hofer AM. Deoxycholic acid activates protein kinase C and phospholipase C via increased Ca2+ entry at plasma membrane. *Gastroenterology*. 2005 Mar;128(3):695-707.

[41] Qiao D, Stratagouleas ED, Martinez JD. Activation and role of mitogen-activated protein kinases in deoxycholic acid-induced apoptosis. *Carcinogenesis*. 2001 Jan;22(1):35-41.

[42] Im E, Martinez JD. Ursodeoxycholic acid (UDCA) can inhibit deoxycholic acid (DCA)-induced apoptosis via modulation of EGFR/Raf-1/ERK signaling in human colon cancer cells. *J Nutr*. 2004 Feb;134(2):483-6.

[43] Jenkins GJ, Harries K, Doak SH, Wilmes A, Griffiths AP, Baxter JN, et al. The bile acid deoxycholic acid (DCA) at neutral pH activates NF-kappaB and induces IL-8 expression in oesophageal cells in vitro. *Carcinogenesis*. 2004 Mar;25(3):317-23.

[44] Glinghammar B, Inoue H, Rafter JJ. Deoxycholic acid causes DNA damage in colonic cells with subsequent induction of caspases, COX-2 promoter activity and the transcription factors NF-kB and AP-1. *Carcinogenesis*. 2002 May;23(5):839-45.

[45] Qin P, Tang X, Elloso MM, Harnish DC. Bile acids induce adhesion molecule expression in endothelial cells through activation of reactive oxygen species, NF-kappaB, and p38. *Am J Physiol Heart Circ Physiol*. 2006 Aug;291(2):H741-7.

[46] Qiao D, Chen W, Stratagoules ED, Martinez JD. Bile acid-induced activation of activator protein-1 requires both extracellular signal-regulated kinase and protein kinase C signaling. *J Biol Chem*. 2000 May 19;275(20):15090-8.

[47] Jenkins GJ, Cronin J, Alhamdani A, Rawat N, D'Souza F, Thomas T, et al. The bile acid deoxycholic acid has a non-linear dose response for DNA damage and possibly NF-kappaB activation in oesophageal cells, with a mechanism of action involving ROS. *Mutagenesis*. 2008 Sep;23(5):399-405.

[48] Araki Y, Katoh T, Ogawa A, Bamba S, Andoh A, Koyama S, et al. Bile acid modulates transepithelial permeability via the generation of reactive oxygen species in the CACO-2 cell line. *Free Radic Biol Med*. 2005 Sep 15;39(6):769-80.

[49] Gonzalez-Mariscal L, Tapia R, Chamorro D. Crosstalk of tight junction components with signaling pathways. *Biochim Biophys Acta*. 2008 Mar;1778(3):729-56.

[50] Leso V, Leggio L, Armuzzi A, Gasbarrini G, Gasbarrini A, Addolorato G. Role of the tumor necrosis factor antagonists in the treatment of inflammatory bowel disease: an update. *Eur J Gastroenterol Hepatol*. 2010 Jul;22(7):779-86.

[51] Hoentjen F, van Bodegraven AA. Safety of anti-tumor necrosis factor therapy in inflammatory bowel disease. *World J Gastroenterol*. 2009 May 7;15(17):2067-73.

[52] Ma TY, Iwamoto GK, Hoa NT, Akotia V, Pedram A, Boivin MA, et al. TNF-alpha-induced increase in intestinal epithelial tight junction permeability requires NF-kappa B activation. *Am J Physiol Gastrointest Liver Physiol*. 2004 Mar;286(3):G367-76.

[53] Graham WV, Wang F, Clayburgh DR, Cheng JX, Yoon B, Wang Y, et al. Tumor necrosis factor-induced long myosin light chain kinase transcription is regulated by differentiation-dependent signaling events. Characterization of the human long myosin light chain kinase promoter. *J Biol Chem*. 2006 Sep 8;281(36):26205-15.

[54] Mashukova A, Wald FA, Salas PJ. Tumor necrosis factor alpha and inflammation disrupt the polarity complex in intestinal epithelial cells by a posttranslational mechanism. *Mol Cell Biol*. 2011 Feb;31(4):756-65.

[55] Shen L, Su L, Turner JR. Mechanisms and functional implications of intestinal barrier defects. *Dig Dis*. 2009;27(4):443-9.

[56] Atreya I, Atreya R, Neurath MF. NF-kappaB in inflammatory bowel disease. *J Intern Med*. 2008 Jun;263(6):591-6.

[57] Danese S, Mantovani A. Inflammatory bowel disease and intestinal cancer: a paradigm of the Yin-Yang interplay between inflammation and cancer. *Oncogene*. 2010 Jun 10;29(23):3313-23.

[58] Baeuerle PA, Baltimore D. I kappa B: a specific inhibitor of the NF-kappa B transcription factor. *Science*. 1988;242:540-6.

[59] Karin M. How NF-kB is activated: the role of the IkB kinase (IKK) complex. *Oncogene*. 1999;18:6867-74.

INDEX

A

B

C

D

E

F

Q

R

S

T

U

V

W

Z